LOOP CHECKING
A TECHNICIAN'S GUIDE

LOOP CHECKING
A TECHNICIAN'S GUIDE

Harley Jeffery

ISA TECHNICIAN SERIES

Copyright © 2005 by ISA – Instrumentation, Systems, and Automation Society
67 Alexander Drive
P.O. Box 12277
Research Triangle Park, NC 27709

All rights reserved.

Printed in the United States of America.
10 9 8 7 6 5 4 3
PRINTED February 2006

ISBN-10: 1-55617-910-3
ISBN-13: 978-1-55617-910-5

No part of this work may be reproduced, stored in a retrieval system, or transmitted in any form or by any means, electronic, mechanical, photocopying, recording or otherwise, without the prior written permission of the publisher.

Notice

The information presented in this publication is for the general education of the reader. Because neither the author nor the publisher has any control over the use of the information by the reader, both the author and the publisher disclaim any and all liability of any kind arising out of such use. The reader is expected to exercise sound professional judgment in using any of the information presented in a particular application.

Additionally, neither the author nor the publisher have investigated or considered the effect of any patents on the ability of the reader to use any of the information in a particular application. The reader is responsible for reviewing any possible patents that may affect any particular use of the information presented.

Any references to commercial products in the work are cited as examples only. Neither the author nor the publisher endorses any referenced commercial product. Any trademarks or tradenames referenced belong to the respective owner of the mark or name. Neither the author nor the publisher makes any representation regarding the availability of any referenced commercial product at any time. The manufacturer's instructions on use of any commercial product must be followed at all times, even if in conflict with the information in this publication.

Library of Congress Cataloging-in-Publication Data

Jeffery, Harley.
  Loop checking :a technician's guide / Harley Jeffery.
    p. cm. -- (ISA technician series)
  Includes bibliographical references.
  ISBN 1-55617-910-3 (pbk.)
  1. Process control--Automation. I. Title. II. Series.
  TS156.8.J44 2005
  629.8'3--dc22

2005001679

To Tamra, John, Chrissy, Ashton, and Sarah

ACKNOWLEDGMENTS

During the past several years, I have met many people who have shared their invaluable and memorable knowledge and insights. *Loop Checking: A Technician's Guide* combines this wisdom with the experiences of control loop device manufacturers, control system engineers, plant I&E technicians, and mentors/educators. Although these contributors are too many to list here, the following people deserve special recognition for having shared their vision and enthusiasm for process controls and/or life experiences: Dr. John Ristroph, Bill Ruple, Charlie Brown, Chick Childress, Dr. Chun Cho, Floyd Jury, Don Lambert, Terry Blevins, Harley and Anne Jeffery, Jr., Dave Thomas, Rev. Cy Mallard, Mel England, Harry Pinder, Bob Zielski, Chris Liakos, Greg McMillan, Stan Weiner, Dr. Lawrence Mann, and Bill Bialkowski. Their expertise and insight are much appreciated.

ABOUT THE AUTHOR

Harley Jeffery has worked in the field of Industrial Process Controls for over 30 years. Beginning his career with Fisher Controls as a valve application engineer, he spent the last 23 years with Control Southern as a control systems engineer. His experience includes the design, implementation, testing, start-up, and ongoing improvement of analog, distributed, and digital control systems. Harley received a BSIE from Louisiana State University in 1974 and completed Monsanto's Electrical and Instrument Engineering School in 1978. Currently, he is a process control consultant and project manager with Control Southern Inc. He can be reached at harley.jeffery@controlsouthern.com.

TABLE OF CONTENTS

PREFACE		xiii
Chapter 1	**Introduction to Loop Checking**	1
	1.1 The Opportunity	2
	1.2 Loop Checking: Introduction	3
	1.3 Process Control Example	10
	1.4 Other Loop Checking Considerations	11
	1.5 Control Loop Design Guidelines	21
Chapter 2	**The Factory Acceptance Test**	25
	2.1 Documentation	26
	2.2 Test Planning	26
	2.3 Performing the FAT	27
	2.4 Process Simulation	38
	2.5 Process Control Example	40
Chapter 3	**Start-up**	47
	3.1 Documentation	47
	3.2 Loop Check Plan	48
	3.3 Checking the Loop	48
	3.4 Process Control Example	51
	3.5 Example Forms	54
	References	57
	Quiz	57
Chapter 4	**Performance Benchmarking**	59
	4.1 Designing the Test	60

	4.2 Performing the Test	62
	4.3 Analyzing and Reporting the Test Results	72
	4.4 Process Control Example	79
	References	87
	Quiz	88
Chapter 5	**Sustaining the Performance**	**89**
	5.1 Maintenance Strategies	89
	5.2 Operator Efficiency	91
	5.3 Where to Start?	91
	5.4 Selecting the Scope of Your Loop Performance Program	91
	5.5 Loop Performance Monitoring/Analyzing	93
	5.6 Performance Reporting	104
	5.7 Loop Performance Program Architecture	105
	5.8 Loop Performance Program Summary	109
	References	110
	Quiz	111
Acronyms		**113**
Appendix A	**Tuning**	**115**
Appendix B	**Answers**	**135**
Index		**139**

PREFACE

During a recent control system checkout prior to start-up, I was surprised at the project engineer's comment that they could have done a better job at planning. The project experienced a successful start-up—due to extensive planning, preparation, and testing. However, some aspect of checkout and start-up could always be improved, especially by more planning. Thus, one of the key objectives of this guide is to pass along examples, hints, and methodologies for planning and implementing "loop checking." Even though the term "loop checking" may only suggest the activity immediately prior to a control system start-up, the foundation for successful start-up begins in the project design and acceptance testing stage, and continues through start-up, performance benchmarking and sustaining the performance. Thus, this guide's chapters are in the following sequence.

This guide covers the main tasks in a typical control system automation project "loop checking" sequence, but does not delve into specific activity details—such as loop design, instrument calibration, loop tuning, etc.—that are covered in many other articles and books. Instead, "loop checking" program elements can be customized for your specific implementation, based on your plant's philosophies and preferences.

Chapter 1 – Introduction to Loop Checking provides a background on the recently increased emphasis on the loop checking process (i.e., the financial paybacks are significant from control system performance). After a brief definition of the control loop elements, a loop checking process flow diagram is provided along with comments on technology improvements in smart field devices. A "Process Control Example," using a "generic" application of boiler drum level control found in many different industry segments and combining feedback with cascade and feedforward control, is included that reinforces the chapter discussions. This feature appears in each chapter. Finally, several general topics to be considered in loop checks are discussed.

Chapter 2 – The Factory Acceptance Test delves into the acceptance testing that takes place before start-up, where planning and verification can save significant time and expense. Although planning is important in all phases, it is particularly useful here for both the vendor and end user to make sure that a clearly understood plan of "who does what" and "what is expected" is in place. Divided into *hardware* and *configuration* loop checks, this chapter presents a "test plan" you may customize for your particular project needs. The use of a process simulation package for testing is discussed, along with the impact of smart instruments. The chapter concludes by applying the test plans to our example process, with some typical forms that might be used or modified to your preference.

Chapter 3 – Start-up discusses the planning and activities to be considered during the start-up loop check. Again, to maintain a team approach, efficient use of everyone's time is a key focal point for planning the start-up loop checks. "Smart" field instruments can have a large impact on this activity; the chapter takes a closer look at loop checking with and without smart instruments. A start-up scenario provides the process control example in Chapter 3.

Chapter 4 – Performance Benchmarking reviews this tool which checks the loop and ensures performance is at the highest level possible, and which can set the benchmark for performance that is monitored over time (Chapter 5). Although performance testing occurs after start-up in this project sequence, the procedure can also be used with any loop/unit operation as a troubleshooting guide when process control problems arise, or as a general approach to improve existing operations. Again, planning is important; items to watch for are suggested and typical forms are included, along with some recommendations on how to perform the checks. The example process is then "benchmarked" using the techniques discussed in this chapter.

Chapter 5 – Sustaining the Performance makes sure you can maintain the high level of performance achieved and benchmarked in Chapter 4. Maintenance strategies show where the performance monitoring fits. Suggestions on starting this program are included along with some in-depth considerations of how the program will address the elements of the loop. The implementation of a performance monitoring program is dependent on the particular control system architecture (single loop controllers, PLCs, DCSs, hybrid control, digital control systems, etc.) that your plant has installed (or is planning to install). Thus, several of these hardware architectures are examined with ideas and options for implementation.

1
INTRODUCTION TO LOOP CHECKING

Process control loops have a major impact on the financial performance of today's manufacturing facilities. It is also recognized that a "good foundation" of the basic regulatory control loop is essential to the success of higher-level "Advanced Process Control (APC)" program (Ref. 6). Thus, it is critical that these facilities' technicians focus on loop checking and performance. For this reason, this guide defines *loop checking* broadly to include control loop performance rather than merely in terms of plant start-up situations as in the traditional definition. Loop checking is also an important component in any plant's continuous improvement planning program insofar as it helps define and reduce the variability of key process parameters on an ongoing basis.

The chapters of *Loop Checking: A Technician's Guide* are arranged to follow a typical automation project from design checkout at the factory acceptance test (FAT) through to an ongoing sustaining loop performance program. The steps of such projects is as follows:

- loop checking basics
- the factory acceptance test (FAT)
- start-up
- performance benchmarking
- sustaining the performance

This guide is intended to discuss general methods and practices that can be applied across many processes or industries. The technician will encounter different plans and programs in his or her own company for addressing loop performance. These will, of course, affect how loop checking is defined and accomplished for the technician's specific environment. However, the instrument technician typically has the best overall knowledge and skills for checking and maintaining control loop performance.

1.1 THE OPPORTUNITY

In today's intensely competitive markets, manufacturers are striving to continually improve manufacturing performance to meet their business needs and goals. Typical business drivers are as follows:

- increased throughput
- increased yield
- increased quality
- minimized waste and off-spec

As we noted, the control loop (and the continual checking of performance) plays a vital role in the plant's financial performance. However, it has been observed that up to 80 percent of all loops are not performing their intended function of reducing the variability that results from the problems caused by the factors shown in Figure 1-1. Such issues as measurement placement and the dead time or process mixing it causes, undersized headers and valves, loop tuning, and control strategy, all affect the loop's ability to accomplish the desired objectives.

FIGURE 1-1
Control Loop Performance Issues

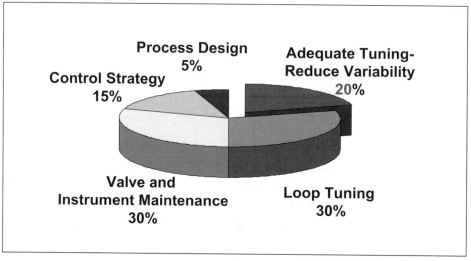

In addition, plant performance studies (such as those summarized in Figure 1-2) have shown that the largest opportunity for reducing costs (1.5%) is provided by field device performance and loop tuning, where loop checking methods can be applied.

FIGURE 1-2
Process Improvement Opportunities

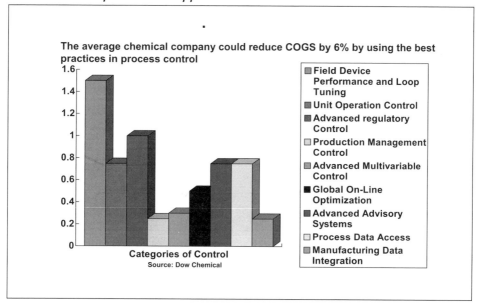

The Control System Technician (CST) can become involved in the performance of the plant's control loops, from the control implementation stage through to the checkout phase and then continuing through start-up, commissioning, and ongoing operations.

1.2 LOOP CHECKING: INTRODUCTION

The following section reviews the components of the control loop and the scope of loop checking.

Defining the Loop

The purpose of control loops has been defined in various ways:

- to force the process to perform in a predetermined, desirable manner. The process may be a flow, pressure, temperature, level, or some other variable in the manufacturing plant (Ref. 3).

- to adapt automatic regulatory procedures to the more efficient manufacture of products or processing of material (Ref. 4).

- to ensure safety, environmental regulation, and profit (Ref. 5).

Basic to any discussion of control loops is "feedback" control. In this control, the loop starts by measuring the process variable (PV). It then compares the PV to the desired value, that is, the set point (SP), and acts on the difference between SP and PV (error) using a control algorithm (typically PID). The loop then outputs to the final control element. The diagrams below indicate that the main elements of the loop are:

- transmitter/sensor (for measuring the PV)
- process controller (with an operator-entered SP and control algorithm)
- final control element (valve/actuator and accessories)

Control system engineers use the block diagram in Figure 1-3 to show the relationships of the control loop elements.

FIGURE 1-3
Feedback Loop Block Diagram

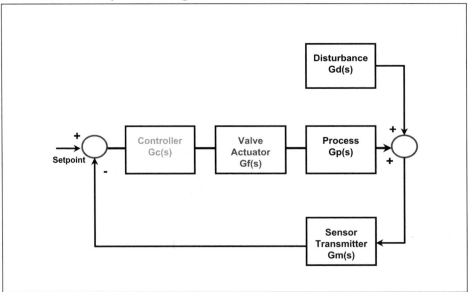

In a more practical view, the block diagram looks like Figure 1-4 below when depicted with hardware for measurement, controller and final control element functions.

FIGURE 1-4
  Control Loop

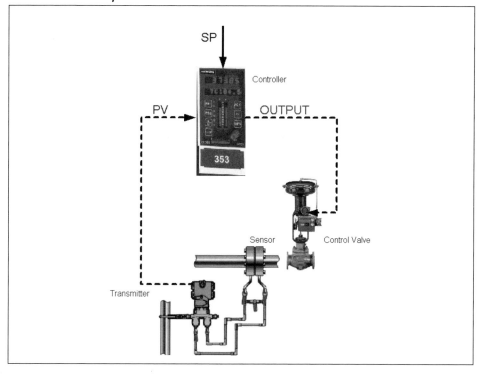

For the purposes of this guide, we'll focus on the single input, single output control loop as depicted in Figures 1-3 and 1-4.

Elements of the Loop

Let's discuss each element in the loop. Although several chapters could be dedicated to each element of the loop (a good resource is Reference 1), we will try to keep the discussion brief and highlight important features for our expanded definition of loop checking, which includes performance.

Sensor / Transmitter

The loop starts here and cannot do a good job unless the measurement is accurate, reproducible (reproducibility is the closeness of agreement of an output for and input approaching from either direction at the same operating conditions over a period of time and is a better number for control and measurement evaluations – see Reference 6). Total Probable Error is another important performance specification that you may use for comparison purposes. Measurement resolution of the signal within the control system is usually not an issue with today's control systems I/O design, but if you configure the loop for large spans (watch for temperature loops), then small changes can go undetected. Of course, proper selection and installation of the sensor and transmitter is critical based on service conditions, accuracy, reproducibility, stability, reliability, and other plant standards. Deadtime and noise introduced by measurement installation location can really hurt the loop performance (the typical controller, proportional, integral and derivative [PID] does not handle deadtime very well). For example, mounting a consistency transmitter so that it is convenient to work on versus placing it near the dilution source can introduce unwanted deadtime, while a sensor installed near the valve outlet instead of upstream of the valve will have excessive noise.

Controller

The controller compares the transmitter measurement (PV) to the operator-entered set point (SP), calculates the difference (error), acts on the error with a PID algorithm and outputs a signal to the valve. Today's control systems all have very capable controllers but you need to be aware of the type PID algorithm that your plant's control system manufacturer has implemented. The two (2) common types are called "classic" and "non-interacting" (Ref. 5). Others have called them "series" and "parallel". There is a difference in how you tune the loop with these two types of controllers. If your plant has just one control system, then your plant standard tuning methods can be used without worrying about this difference. However, as plant purchases come from different vendors/ OEM's, different control system types are employed (e.g., programmable logic controller [PLC] vs. distributed control system [DCS]). You need to pay attention when tuning the different controllers to make sure the right tuning methods are applied. The microprocessor-based systems have also introduced us to configurable loop scan (execution) times, which can also be critical to loop performance. You'll want to make sure your controller is executing fast enough for the process dynamics. Table 1-1 suggests a starting point for some typical measurement types.

Final Control Element

The final control element takes the signal from the controller and attempts to position the flow controlling mechanism to this signal. There are various types of final control elements and some have better performance in terms of "positioning" the device. Final control elements can be variable speed drives for pumps or fans, dampers/louvers, heater

TABLE 1-1.
Typical Scan Times

MEASUREMENT TYPE	SCAN RATE RANGE (SEC)
Pressure	0.25 – 2.0s
Flow	0.25 – 2.0s
Temperature	1.0 – 15.0s
Level	1.0 – 5.0s
Conductivity	0.5 – 2.0s
Consistency	1.0s
Analyzer (gas)	1.0s
pH	0.25 – 5.0s
Average Torque	0.5 – 1.0s
Speed	0.25 – 1.0s
Current and Other Electrical Measurements	1.0s
Analog Output	1.0s

controls, but the most common is the control valve. The valve receives the most attention in the loop check because it receives an electrical signal from the controller (i.e., 4–20 mA current or digital value on a bus), converts the electronic signal to a pneumatic signal that must then drive an actuating device to a precise location. We'll talk more about valve and loop performance in later chapters but you'll be hard pressed to beat a sliding stem valve with spring-and-diaphragm actuator and a two-stage positioner for performance. In addition to the controller performance enemies of deadtime and noise mentioned previously, the valve could also introduce non-linearities and deadband into the loop – neither of which is good for the PID controller. In receiving an electronic signal and converting it to a valve plug/ball/disk position in the pipe, various sources of non-linearity and deadband can build up. Friction from seals and packing, backlash of mechanical parts, relay dead zones, shaft windup can keep the valve from maintaining the signal required by the control system. Proper valve sizing and selection of valve characteristic can help linearize the flow response to controller output changes – again very important to how the PID can perform.

Other Loop Types

In addition to feedback control, technicians will encounter several other control strategies when performing loop checking, such as cascade, ratio, and feedforward control. The process control example in this chapter briefly discusses these techniques but the same basics apply to verify the input/control and design/output of the loop.

In some plants, the term "loop" may also include other control system functions such as Analog Indicate Only, Motor Start/Stops, On-off Valves, Discrete Input/Output type control functions. A detailed discussion of these functions is not included in this guide although you could easily expand the methods and techniques to include them in your plant's loop check plan.

There are several excellent resources that go into more depth on each of the elements of the loop. Vendor literature and application papers are good sources of information as are a variety of industry publications (e.g., ISA, TAPPI, etc. – Ref. 5 and 6).

Loop Checking

Some think of loop checking as a process to confirm that the components of the loop are wired correctly and is typically something done prior to start-up. However, due to factors described in the introduction above, the loop check's scope has expanded to also include tests to confirm that it is "operating as designed" and then to ongoing programs for benchmarking and monitoring performance. The block diagram in Figure 1-5 illustrates the components of this expanded loop checking process.

This process starts when the instruments are received at the plant site. It continues through installation and start-up and into the ongoing plant operation. In addition, the control system should perform the intended function properly. This includes verifying the transmitter's process variable (PV) for display to the operator, for use in the control strategy, and for historical trending. This verification testing prior to start-up is known as the factory acceptance test (FAT) which, as an option, can be duplicated at site with the actual hardware and software installed, termed site acceptance test (SAT). Further discussed in Chapter 2, the FAT can be performed prior to shipment of the hardware or in parallel with the hardware installation at site if the overall start-up schedule is compressed.

Once the technician checks the control strategy to verify that the expected output to the final control element is produced, the loop can then be commissioned and start-up can proceed.

Finally, the loop check can include defining the loop performance benchmark and providing a method for monitoring the performance over time.

FIGURE 1-5
Loop Checking Process

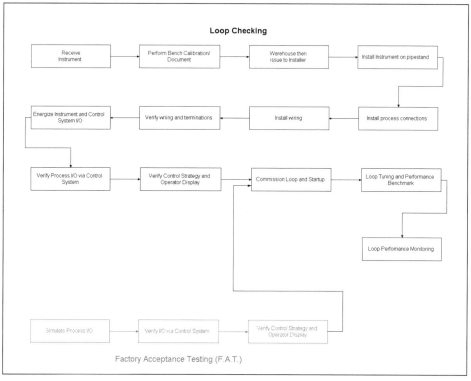

Because of the scope of this manual, we will not cover the receiving, calibration, or installation aspects of the loop check (see Ref. 7). Instead, we will focus more on the verification, start-up, and monitoring phases.

Technology Improvements

Recent developments in "smart" instruments, I/O buses, and software products have made possible increased flexibility and productivity in the loop checking process. Such smart digital technology makes it possible to access and use new types of information that were not available from the analog 4-to-20-mA transmitter and valve. For example, HART® and FOUNDATION™ Fieldbus devices give the technician access to significant amounts of diagnostic, calibration, and performance information, not only about the device but the process as well. Software packages store the extensive data for analysis, future reference, and regulatory agency documentation as well as include sophisticated troubleshooting assistance. As a result, fewer people and less time are required to perform the conventional tasks involved in the checkout process. However, along with the new technology, new tool and training requirements must be followed in order for plants to be able to capitalize on the advertised benefits. Subsequent chapters of this guide will discuss the impact of smart technology on the various phases of loop checking.

1.3 PROCESS CONTROL EXAMPLE

The following example of boiler drum level, feedwater, and steam flow (three-element control) will be used throughout this guide to illustrate the loop check (see Figure 1-6). By combining feedback, cascade, and feedforward control techniques, we can cover several aspects of loop checking simultaneously.

The control objective in this example is to maintain drum level at an operator-entered set point within close tolerances throughout the boiler's operating range. This is achieved by controlling the feedwater inlet flow, with assistance from steam flow to compensate for load disturbances. Close control to the level set point is desired because of process equipment safety issues. These include concerns over water carryover into steam lines resulting from high drum levels and the potential for boiler tube damage as a result of low water levels.

FIGURE 1-6
Example of Boiler Drum Process Control Loops

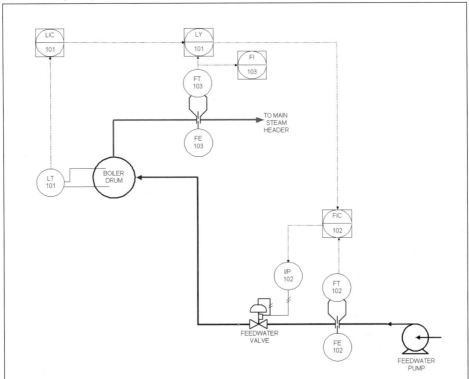

Measurement. For the purposes of this example, the flow and level measurements are accomplished using differential pressure devices. The feedwater and steam flows utilize orifice plates to develop a differential pressure that is proportional to the square of the flow. The drum level transmitter reads a differential pressure signal between the water and steam in the drum and an ambient water column.

Control System. We assume either a distributed control system (DCS) or programmable logic controller (PLC) control system are being used. All controller algorithms are proportional-integral-derivative (PID). The control strategy involves the following loop types:

- Feedback Control—The output of the feedwater flow loop controls the feedwater valve in accordance with a set point cascaded from the drum level loop and compared to the feedwater flow measurement. This closed loop control of the feedwater flow in a cascade system allows the feedwater flow loop to correct for any disturbances in the feedwater flow before those disturbances affect the drum level.

- Cascade Control—The drum level controller (primary/master loop) compares the drum level measurement to the operator set point and outputs to the set point of the feedwater flow controller (secondary/slave loop).

- Feedforward Control—The feedforward action is accomplished by summing the steam flow measurement together with the corrective output from the drum level controller. This signal is then used as the set point for the feedwater flow loop.

Final Control Element. This example assumes a pneumatic-operated, sliding stem control valve with a digital valve controller for manipulating the feedwater.

1.4 OTHER LOOP CHECKING CONSIDERATIONS

Summarized below are some other considerations for designing, implementing, checking, and benchmarking control loops. These guidelines are based on the philosophy that control systems should minimize product variability and improve the overall efficiency of an operation. Instrument and control system manufacturer's installation and maintenance guidelines should also be followed.

Keeping Score

How do we know when a loop is performing well? We need some kind of "measure" that will identify the loop performance. As we'll discuss further in Chapters 4 and 5, many different methods are used to measure the performance of the loop. You'll need to discuss with your performance team which approach is best for your plant.

Some people have mentioned the following as performance measurements (Ref. 11):

- The plant didn't blow up
- The process measurements stay close enough to the set point
- They say it's fine, and you can go home now

Seriously, there are a host of mathematical techniques to measure how the loop is performing. This section briefly discusses a few of these measurements that have been used in process control applications.

One method is to calculate the "variability" of the loop so that we can get an idea of the relative performance compared to other loops. Variability can be defined as two times the standard deviation divided by the mean. Let's look at each of these terms. In Figure 1-7, as the process variable (PV) changes, there will be a band of values, which typically form a distribution or spread (2σ) about some "mean" (average) value.

FIGURE 1-7
Loop variability

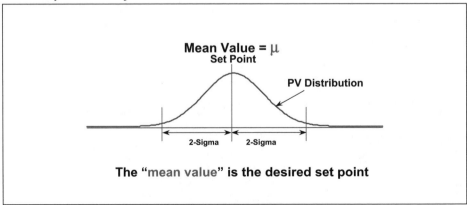

Loop Checking

Standard Deviation (Sigma), σ, is a statistically derived parameter that describes the "spread" of data about the mean value. The larger the value of σ, the greater the spread. The "mean" is simply the average of the values, and in loop analysis, this typically calculates closely to the set point.

Statistics again tell us that the spread of data represented by 1-Sigma will encompass about 68% of the total data values while 3-Sigma will get 99.73%. See Figure 1-8 for a representation. As sort of an industry benchmark, we've settled on the 2-Sigma value to use for our process control variability calculation.

FIGURE 1-8
Picking the Data Spread

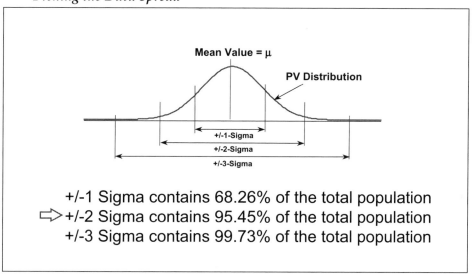

Thus, when you divide 2-Sigma by the mean, you now will have a value in % that will then allow you to compare loops that, for example, control flow (perhaps measured in gallons per minute or GPM) with level loops (engineering units in feet). Of course, some loops such as the secondary cascade loops or certain level loops are designed to absorb some of the process variability and thus a higher variability is fine.

Consider the following example:

> Figure 1-9 shows data on a recorder from a flow loop transmitter while the loop is in MANUAL with random disturbances causing the flow to vary. By using commercially available spreadsheet tools or loop analysis software, the statistics of two times standard

deviations (2σ, Sigma) of 26.6 GPM, mean (μ, average) of 603 GPM and resulting variability (2s / μ) of 4.4% can be calculated.

FIGURE 1-9
Flow Example with Statistics

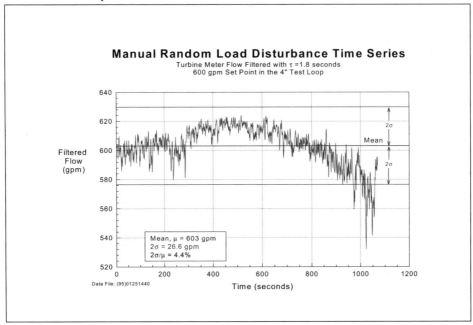

You would hope that by placing the loop to automatic and repeating the test in Figure 1-9 that the calculated variability would be reduced. If it's not, then read on in Chapters 4 and 5 for loop troubleshooting techniques that can help reduce variability.

So, what is "good variability" for a loop? The table below suggests a variability range for key loops in your process as a rule of thumb.

TABLE 1-2.
Variability Rating

VARIABILITY	RATING
Less than 0.5%	Excellent
Less than 1.0%	Very Good
Less than 2.0%	Good
Less than 5.0%	Fair
Less than 10%	Unacceptable
Greater than 10%	Terrible

In addition, there are a number of software packages on the market that are very capable of evaluating control loop performance. The following examples are from vendor information describing their evaluation techniques:

Variability Index - a comparison of current loop operation to minimum variance control. For example, if the variability index is zero(0), then the control loop performance can not be improved. If the variability index is 100, then the loop is doing nothing to reduce process variation (i.e., its performance isn't any better than if it were on manual). Variability index is not fooled by noise or load disturbances (i.e., it is a true indication of how far off current performance is). A variability index of zero(0), for example, indicates no improvement is possible; a variability index of 100 indicates that the loop is providing no benefit in reducing variability.

Harris Index - a performance measure typically discussed by academics is the Harris Index. The Harris Index looks at the error signal, which is the process variable or measurement minus the set point. The Harris measures the ratio between the error variance and the variance achievable by a minimum variance controller. The larger the value, the poorer the performance of the loop. The Harris Index calculation results in a number between 1 and infinity. A value of 1 is perfection or minimum variance control. Larger numbers might be considered worse.

$$\frac{\text{Current Variance}}{\text{Minimum Variance}}$$

- 1 = perfect
- larger = poorer performance

Another form of the Harris Index is the CLPA or Closed Loop Performance Assessment. It is simply the Harris Index, normalized to be between 0 and 1:

CLPA = 1 - 1/(Harris Index)

With the CLPA:

- 0 = perfect control
- 1 = poorest control

ExperTune Index - measures how well a control loop responds to process upsets. The index uses a process model combined with current and optimal PID tuning values. The index can be found by simulating the response of the control loop to a load upset with both sets of tuning values. The simulation provides the data to calculate the integrated absolute error (IAE) between the set point and process variable for each case. With the IAE for each case, the comparison can be made. The ExperTune index is:

$$100 \times \frac{\text{Current IAE} - \text{Optimal IAE}}{\text{Current IAE}}$$

With the ExperTune Index:

- 0 = perfect control

- larger = % performance improvement possible

The metric is unitless and provides a meaningful comparison between loops. This metric will catch those loops that have been de-tuned.

Based on performance demographics of twenty-six thousand PID controllers collected over the last two years across a large cross sample of continuous process industries, an algorithm combining a minimum variance benchmark and an oscillation metric tuned for each measurement type (flow, pressure, level, and so on) was used to classify performance of each controller into one of five performance categories. These classifications were refined through extensive validation and industry feedback to reflect controller performance relative to practical expectations for each measurement type. Unacceptably sluggish or oscillatory controllers are generally classified as either "fair" or "poor" while controllers with minor performance deviations are classified as "acceptable" or "excellent".

Loop Nonlinearities and Deadtime – The Bad Guys

By far the most widely used control algorithm in process control loops is the proportional – integral - derivative (PID) controller. Although there are many techniques to "tuning" the PID settings (which is beyond the scope of this guide) for best response to process upsets, a basic underlying assumption is that the process response is approximately linear with little change in installed/process gain and minimal deadtime. The more nonlinearities and deadtime that creep into a loop, the more the loop has to be de-tuned and thus may not be able to meet your objectives.

In fact, nonlinearities and deadtime can even cause a loop to amplify the disturbances so that control is worse in automatic than if the loop was in manual! Obviously, being able to identify and fix/minimize nonlinearities and deadtime is key to loop performance.

The above mentioned guidelines discussed some methods for reducing nonlinearities and deadtime from a design standpoint in the loop checking process but you'll also want to be aware of other sources. For example, control valves introduce friction, backlash, shaft-windup, and relay dead zones while transmitters can have damping filter and sample time issues and the control system also can introduce filter and control strategy nonlinearities. To help minimize these problems, purchase high-performance valves/positioners and transmitters and then utilize maintenance programs to sustain this performance over time.

Control valve sizing and selection can also play a large part in the overall loop linearity.

As mentioned previously, we want the overall response of the process—that is, when the valve moves and the process responds—to be as linear and as constant gain as possible over the operating range. This overall process response is called the "installed characteristic" of the loop. Let's look at how the installed gain is important. For example, in looking at the pump and system A curve in Figure 1-10, as the flow through the system increases, the outlet pressure, or head, of the pump drops off while the system losses through pipe tees, elbows check valves, and so on increases. The throttling control valve absorbs the difference between the two pressures. Notice the difference in the pump and system B curve shown in Figure 1-11, where the pressure drop in which the valve must throttle does not change as much as it does in system A.

So, how do you select valves to handle these varying pump/system applications and still provide a linear response for the PID loop? The control valve "trim" (the internal parts of the valve that control the flow passage—for example, plug, seat ring, and cage) is engineered to fit the application by shaping the plug or cage windows to provide what's called the "inherent characteristic." You select the trim with the inherent characteristic that will best linearize the pump/system curve. Most valve vendors develop a table/curve by testing each valve that shows the flow resulting from valve movement with a constant pressure drop. In general, the three most common "inherent characteristics" that are available to choose from include "equal percentage," "linear" and "quick opening" (see Figure 1-12).

FIGURE 1-10
Pump and System A Curve (Ref. 12)

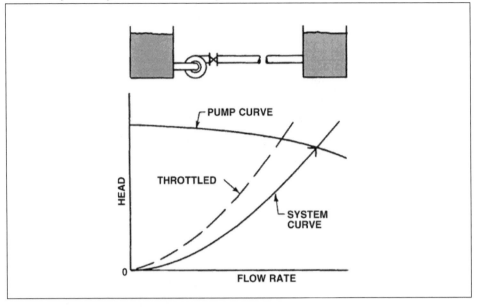

FIGURE 1-11
Pump and System B Curve (Ref. 12)

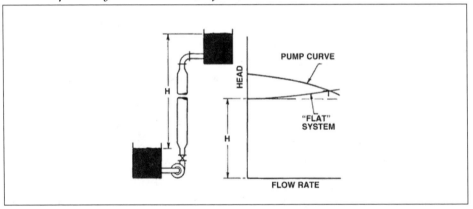

Let's see how this works. For the pump and system curve A where the throttling pressure drop across the valve decreases as flow increases, the selection of an "equal percentage" inherent characteristic would be best because the reduction in pressure drop is "canceled out" by the increasing flow area of the valve. The net result is that the "installed characteristic" now becomes more of a linear response for the range of flowing conditions – good news for the PID loop and your chance to tune for tight control. Similarly, for a header pressure control example, the inlet and outlet pressures do not change much over the flowing range and a "linear inherent characteristic" would be your best choice.

Loop Checking

You're not out of the woods yet. Even though you know the inherent valve characteristic, you must also take into account valve type (rotary, sliding stem) to get the true picture of what you're up against for loop checking.

FIGURE 1-12
Control Valve Inherent Characteristic

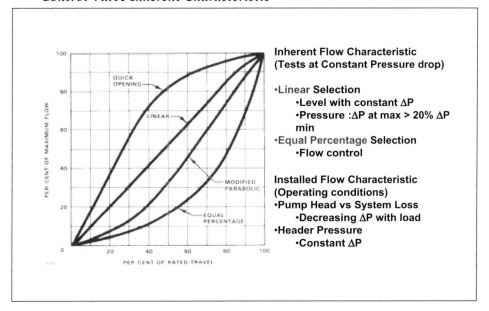

For example, in Figure 1-13, a butterfly style valve is tested in a flow loop for the installed gain the loop will experience. Notice that only a small part (from about a 20 to 40 degree opening provides a linear response with a process gain that is within a desired range (called the EnTech Gain Specification). However, compare this to Figure 1-14 for a sliding stem control valve, which shows that the gain is within spec over a wider operating range. By looking at the valve constructions/flow paths, you can see why the butterfly valve has the narrower range.

FIGURE 1-13
Butterfly Valve Installed Gain

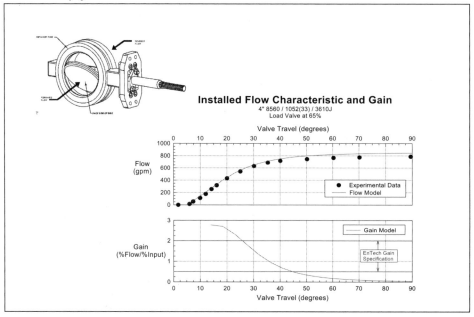

FIGURE 1-14
Sliding Stem Valve Installed Gain

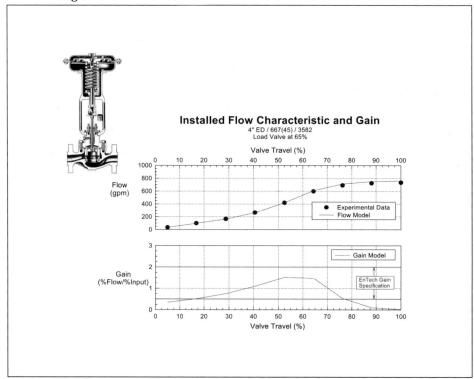

TABLE 1-3.
Valve Type – Control Range Summary

STYLE	CONTROL RANGE (% OF TRAVEL)
Globe/Sliding Stem	60
Vee-ball/Rotary	55
Eccentric ball/Rotary	38
Butterfly	22

Now you can see why the oversized valve trying to throttle near the seat is not going to perform very well. Also, you can see why there is a significant difference at what the top-end control point should be. Imagine trying to tune the loop for a butterfly valve that has to operate near 70 degrees open for one product and 30 degrees for another—probably job security for your loop tuning people. Table 1-3 gives a summary of the different types of control valves and approximate control ranges. There are workarounds such as output signal characterization to ease some of the installed curve nonlinearities or gain scheduling in your controller but spending the time and money to install the correct size and style valve will provide the best solution for the loop checking performance.

1.5 CONTROL LOOP DESIGN GUIDELINES

In this section we summarize eleven guidelines that technicians should consider when designing, implementing, checking, and benchmarking control loops. These guidelines are based on the philosophy that control systems should minimize product variability and improve the overall efficiency of an operation. Technicians should also follow instrument and control system manufacturers' installation and maintenance guidelines. We recommend the following guidelines:

1. Dead time or time delay as seen by the control loop should be minimized wherever possible. This means that (a) measurement devices should be located as close as practically possible to the control device without effecting their measurement characteristics, (b) long instrument tubing runs should be avoided, (c) transmitters with long processing delays should be avoided, and so on.

2. Process transport delay should be minimized.

3. Control devices (valves, air motors, damper drives, etc.) should exhibit repeatable dynamics, with virtually no nonlinearities over the

process's complete operating range. Their speed of response should be reasonable for the application.

4. The control loop's installed process gain from control device to measurement device should nominally be about 1-%span/%output over the process's complete operating range. Process gains in the range of 0.5 to 2.0-%span/%output are acceptable.

5. Filtering of the control loop measurement should be kept to a minimum, and it should be at least five to ten times less than the closed-loop time constant of the loop.

6. Only linear control algorithms should be used. Control loop nonlinearities such as control dead bands, error dead bands, error characterization, and the like, should not be used.

7. The control loop should be tuned using nonoscillatory tuning techniques.

8. Process areas should be tuned in a coordinated manner to minimize loop interaction and disturbances to processes that rely on the ratioing of ingredients or raw materials.

9. The control loop should be stable over the process's complete operating range.

10. The outer loop of a cascaded loop structure should be tuned five to ten times slower than the inner loop.

11. The less critical loop of a set of interacting loops should be tuned five to ten times slower than the more critical loop. (Ref. 8)

REFERENCES

1. Lipták, Béla. *Instrument Engineers' Handbook, 3rd Edition – Process Control, Vol. 2*, (CRC Press/ISA – The Instrumentation, Systems, and Automation Society, 1995).

2. Fitzgerald, Bill. *Control Valves for the Chemical Process Industries* (McGraw-Hill, 1995).

3. Lloyd, Sheldon G. and Gerald D. Anderson. *Industrial Process Control* (Fisher Controls, 1971).

4. Hughes, Thomas A. *Measurement and Control Basics, 3rd Edition* (ISA – The Instrumentation, Systems, and Automation Society, 2002).

5. Sell, Nancy J. *Process Control Fundamentals* (TAPPI Press, 1995).

6. Blevins, Terrence L., McMillan, Gregory K., Wojsznis, Willy K., Brown, Michael W., *Advanced Control Unleashed*, ISA – The Instrumentation, Systems, and Automation Society, 2003.

7. Harris, Diane. *Start-Up: A Technician's Guide*, ISA – The Instrumentation, Systems, and Automation Society, 2001.

8. Bialkowski, W. L. *Process Control for Engineers*, Emerson Process Management, EnTech Performance Group, 2001.

9. Gerry, John. "Performance Measurement – The Rest of the Story." Included with permission of ExperTune Inc. (c) 2003-2004 ExperTune Inc.

10. Desborough, L. D. and R. M. Miller. "Increasing Customer Value of Industrial Control Performance Monitoring—Honeywell's Experience," Proc. CPC VI, 2001.

11. Shaw, John A. *PID Algorithms and Tuning Methods*. n.p., n.d.

12. ITT-Goulds Pumps. *Centrifugal Pump Fundamentals – System Curves*. ITT-Goulds Pumps, n.d.

QUIZ

1. Why would a loop checking program be important to the plant manager/profit center manager? The project manager/engineer? The control system technician? The process engineer?

2. The typical plant can realize the best return on investment by which of the following programs? a) Advanced Control Packages b) Process Design Improvements c) Field Device Performance and Loop Tuning d) Process Data Access and Trending

3. What are the basic elements of the control loop? Which loop element would be most susceptible to long-term performance degradation and thus a candidate for specific device monitoring?

4. When does the loop checking process start? When does it end?

2

THE FACTORY ACCEPTANCE TEST

The factory acceptance test (FAT) is the phase of an automation project in which the control system hardware is staged at the vendor's facility so the configuration is complete and ready to check as an integrated system. The main objectives are to verify that the system is setup correctly and will operate as expected from both the hardware and configuration software points of view *before* it ships to the site, plus the chance to do some training on the new system. With fast-changing technology and variety of control system approaches to loop implementation (e.g., fieldbus, distributed control systems, single loop controllers), the FAT also becomes important since this may be the first time the exact combination of devices are brought together to function as a system.

Figure 1-5 in Chapter 1 shows that this testing occurs prior to the commissioning and start-up phase. Some plants opt for a repeat of the FAT at site with the hardware and field wiring installed, which is called the site acceptance test (SAT).

The loop checking at the FAT can save significant time and effort during the commissioning and start-up at the plant's site because it will find and correct any errors in the software. Then, any troubleshooting effort performed at start-up can focus on field devices and external issues "outside" of the control system's screw terminals. Usually there are more resources and time to fix things at FAT versus at start-up and certainly less pressure. As a side benefit, some plants have used FAT as a training opportunity for operators and maintenance personnel.

Typically, the control system has some sort of simulation (hardware and/or software) that will allow a realistic check of the control strategies against the specifications. The FAT can also include hardware and I/O checks to assure proper system communications. Also, power wiring and grounding, panel layout and bills of material are verified. However, this chapter will only cover the control loop check portion in detail.

2.1 DOCUMENTATION

The first thing the technician needs is a set of documentation on the process to test against. This can be the hardware specification sheets, detailed design specifications, SAMA (Scientific Apparatus Makers Association) drawings, loop sheets, P&IDs, and any control strategy write-ups that are part of the design effort (see Ref. 1, 2, and 3). Usually, the vendor has developed documents with the plant that describe the loop's objectives and functionality. Also, at this point in the project, the configuration of the control system should be complete, and copies of the system's documentation are helpful for verifying the test procedures.

2.2 TEST PLANNING

A thoroughly thought-out test plan will ensure that the plant and technicians get the most out of the time spent testing the system and will pay off at site during start-up. Thus, by checking the I/O from the screw terminals into the control strategy/loop and to the operator display along with any interlocks, all operating a "live (simulated)" process, most system problems are resolved and you should be able to focus any troubleshooting at start-up on field devices or wiring.

The methods for loop check in the FAT can differ depending on the complexity of the checkout desired. For example, a simple test plan may be to just input various signals at the I/O card and watch the variable change on the operator interface without any process simulation. Outputs are tested in a similar manner for observation and check. This method does not usually cover complex control strategies, interlocks, and sequence control unless a large investment is made in signal generators, meters, wiring, and other test equipment.

A more detailed test plan would be to use a PC-based simulation of the process and controller's I/O subsystem. The control system vendors typically have developed these products for their systems where the process controller "talks" to the PC instead of the I/O system and signal generators/meters. Tracing the signals to and from the I/O card channel and the operator display is now done on the PC monitor. The overall checkout time is usually quite a bit faster than the hard-wired method.

As with any plan, the FAT team should have a list of the loop tests to be performed, any special resources required, setup values/conditions, test procedures, and finally a definition of the successful test and how to document the results.

2.3 PERFORMING THE FAT

With the FAT plan developed, you're now ready to do some loop checking (along with the rest of the control system). The following is a brief outline of a possible test based on a "tiered" approach that starts at the "bottom" of the control loop with the I/O channels and progresses up through the basic loops to more complex, interacting loops maybe with interlocks and then to sequence logic and any batch control that may be required. Again, the batch and sequential loop tests are not included in this document.

Hardware and I/O Check

After a kickoff meeting with the vendor to review readiness of the system for test, schedules, team assignments and responsibilities, and to review special test equipment needs and any new issues, the first day's activities generally start with checking the hardware against the purchase order bill of material and any system layout design drawings. Once you're satisfied that the hardware is complete, the next step would be to check the system using the diagnostics package provided with the control system. Typically, this will show that the I/O cards (by location and type) are talking through the communication system devices to operator workstations, historian devices, gateways, and so on, and that there are no hardware diagnostic errors. Although this next test is duplicated when the system is in place (see the next section on loop checking), some companies like to input actual signals to "sample number" of I/O channels and see that the values are communicated to the proper devices. A similar procedure is used with outputs by driving the output channel from the operator workstation and verifying (with a meter and diagnostics) that the proper value is set.

This process not only builds confidence with the checkout team that the basic system tasks will start up in working order once the system is received at the site but also sets the foundation for loop checks to know that the I/O signals are correct. It also provides the team with some training on troubleshooting the new system with the knowledgeable vendor present. Depending on team size and level, we have seen parallel activity at this point such that while the complex controls are verified with the PC-based simulation, other test team members can be checking the I/O and communications to a point where they are satisfied that if a problem occurs during start-up, it is probably "outside" of the control system screw terminals.

The overall objective in the hardware FAT is to verify that the system bill of material is complete and installed correctly and that the basic system functions such as communications and diagnostics are operating with no problems. Some companies have spent time to track FAT deficiencies over several projects to use in quality improvement processes (Ref. 4). The following is an example of the test plan topics that you might use as a starting point for your hardware FAT plan (the example assumes you are using a digital control system but it could also be used for DCSs, PLCs, single-loop panel-based systems, and so on, with some modification).

Section 1 – Introduction

1.1 Define the purpose of your test and expected results. This may just be a simple statement that you want to verify the system hardware against the contract document or add any general statements about additional tests.

1.2 Scope – add some detail to the system scope and test to be run.

1.3 System Drawings/Specifications – define the applicable drawings and specifications that should be on-hand at the test.

1.4 Definitions – add any specific definitions of terms or abbreviations that will be used in the plan.

Section 2 – Test Method Instructions

2.1 General – note the forms to be used and how the test personnel are to indicate test items as complete, unsuccessful – rework required, open for resolution, and so on.

2.2 Detail reporting instructions on how the deficiencies should be categorized, what the corrective action should be and expected result, and how to verify the result.

Section 3 – Bill of Material (BOM)

3.1 Purpose – general statement as to how you will check the hardware pieces – usually a document or drawing is listed as the reference.

3.2 Test Equipment – list any test equipment required.

3.3 Procedures – a form-based procedure or checklist is helpful that can be documented per the Section 2 methods. Listing each BOM item for check-off is typical, inspect for damage, enclosures are per spec.

Section 4 – System Components

4.1 Purpose – general statement to verify that system components are installed and identified correctly.

4.2 Test Equipment – list any test equipment required.

4.3 Procedures – a form-based procedure or checklist is helpful that can be documented per the Section 2 methods. Check for labels and loose wires, I/O cards in the right slot and properly seated.

Section 5 – Grounding and Wire Shielding

5.1 Purpose – general statement to check ground integrity.

5.2 Test Equipment – list any test equipment required (multimeter, for example).

5.3 Procedures – a form-based procedure or checklist is helpful that can be documented per the Section 2 methods. Typically vendors have resistance specs for ground/shield circuits that should be verified and documented.

Section 6 – AC / DC Power

6.1 Purpose – general statement to AC/DC power distribution and voltage levels.

6.2 Test Equipment – list any test equipment required (multimeter, for example).

6.3 Procedures – a form-based procedure or checklist is helpful that can be documented per the Section 2 methods. Verify wiring to breakers, terminal strips and trace to as-powered devices, such as power supplies, cabinet lighting, outlets, and so on. Same for DC powered devices. Measure voltages at the devices to verify proper voltage levels per vendor specs. Test redundant power design by shutting down primary then secondary AC feeds and DC supplies to verify the system continues to function as desired.

Section 7 – I/O Check (optional)

7.1 Purpose – general statement to verify that the I/O is properly communicating through the controller, communication network, and to the operator interface.

7.2 Test Equipment – list any test equipment required (e.g., multimeters, signal generators, jumpers).

7.3 Procedures – a form-based procedure or checklist is helpful that can be documented per the Section 2 methods. Using control system diagnostics, verify that all internal "health" checks are good. With diagnostics or test configurations, set/measure signals to each I/O card and verify at the operator interface. Test any redundancy features such as controllers and communication networks to verify proper switchover.

After this test procedure is completed and accepted - declare victory and sign off the system as ready for configuration checks!

Loop Checking

Once the basic I/O channels and hardware are verified, it is time to start the loop checks. A detailed plan with procedures and checklists will help assure a successful acceptance test and a smoother start-up. It is interesting to note that a company tracking deficiencies in loop checks found the most problems with the configuration aspects of the operator interface piece of loop checks (Ref. 4).

For this guide, it is difficult to define the exact loop control strategy check portion of FAT / SAT due to widely varying methods of configuring the control system. Many organizations spend significant resources defining configuration standards so that using proven methods that are copied from project to project with the associated testing and documentation can reduce project costs. However, we'll attempt to provide some general comments and suggestions.

Focusing on the control loop, there are some typical functions that are encountered on a routine basis and should be described in your test procedures in order to make sure the loop is completely verified. For example:

1. The raw input to the loop is checked along with any calculations or modifications to the signal such as square root, filtering, or compensation.

2. The set point, mode, and output are checked as working in the desired manner. Any loop interlocks, such as tracking or mode change logic, should then be checked.

3. Finally, interaction with other loops is verified if the loop is part of a larger control strategy (see the process control example of drum level and feedwater flow).

As with the hardware and I/O check, you can see where a PC-based simulation system can reduce time requirements due to the ability to quickly make changes to several I/O signals that might be part of the strategy. Also, at the loop-level check, it is very helpful to know how the loop will perform in real process conditions. The PC-based simulation packages also provide for process dynamics (such as level, flow, and temperature) to be included in the signal sent to the loop, which gives a more thorough testing of the loop. Loops can then be placed in automatic and overall control strategies can be checked in conditions that closely follow actual plant conditions.

Figure 2-1 shows the "flow" of a project-testing plan that includes "database, control, and display" testing. This "tiered" approach starts with the basic analog and discrete elements such as indicators, switches and condition/interlock logic and then progresses to basic analog/discrete control loops for verification prior to moving into the complex loop or sequencing logic. As a general rule, do not check any area without first verifying that all preceding areas are accurate and complete. The loop test areas are shown in Figure 2-1 and described in the following subsections.

FIGURE 2-1.
Checkout Flow Chart

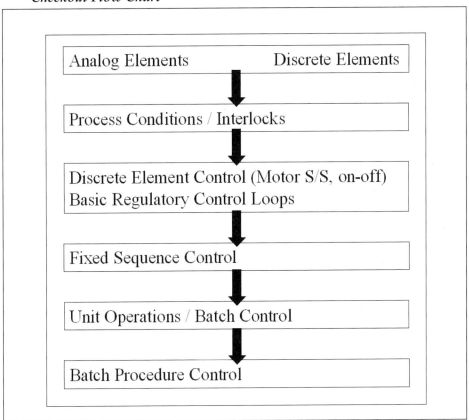

The following test plan is suggested as a starting point for your configuration FAT plan (this example assumes a digital control system but could be used for DCSs, PLCs, single-loop panel-based systems, and so on, with some modification). Example testing forms have been included in the process control example at the end of the chapter. Fixed sequence and batch control guidelines are outside the scope of this guide.

Section 1 – Introduction

1.1 Define the purpose of your test and expected results.

1.2 Scope – include the process areas that are included, number of configuration modules/displays, etc., and testing architecture (PC or signal generators).

1.3 System Drawings/Specifications – define the applicable drawings and specifications that should be on-hand at the test.

1.4 Definitions – add any specific definitions of terms or abbreviations that will be used in the plan.

Section 2 – Test Method Instructions

2.1 General – note the forms to be used and how the test personnel are to indicate test items as complete, unsuccessful – rework required, open for resolution, and so on.

2.2 Detail Reporting Instructions on how the deficiencies should be categorized, what the corrective action should be and expected result, and how to verify the result.

Section 3 – Base Level Analog and Discrete Elements

3.1 Purpose – general statement to describe the base level I/O and module functionality of the simple analog and discrete elements and applicable drawings, specs and standards.

3.2 Test Equipment – list any test equipment required (e.g., signal generators, multimeters)

3.3 Procedures – a form-based procedure or checklist is helpful that can be documented per the Section 2 methods:

Analog Element Testing

- ANALOG INPUT POINTS
    - access field
    - process variable field
    - bar graphs for level indication
    - engineering units field
    - alarm window (where appropriate)
- ANALOG OUTPUT POINTS
    - access field
    - set point field
    - engineering units field
    - alarm window (where appropriate)

Discrete Element Testing

- DISCRETE INPUT POINTS
    - access field
    - alarm window (where appropriate)
    - status window (where appropriate)
- DISCRETE OUTPUT POINTS
    - access field
    - set point field
    - alarm window

Section 4 – Basic Regulatory Control Loop Elements

4.1 Purpose – general statement to describe the basic control functions of the control loop and applicable drawings, specs, and standards.

4.2 Test Equipment – list any test equipment required (e.g., signal generators, multimeters)

4.3 Procedures – a form-based procedure or checklist is helpful that can be documented per the Section 2 methods:

Loop Element Testing

- LOOP POINTS
    - access field
    - process variable field
    - set point field
    - engineering units field
    - alarm window

- mode flag
- implied valve position field
- actual valve position field (when available)
- loop status field
- interlock flag

- PROCESS CONDITIONS
    - all necessary variables configured
    - deadband functionality
    - verbiage and color expressions on interlock/help displays

- MODE LOGIC INTERLOCKS
    - mode interlock conditions match those in the detail design document
    - tag color expressed properly on interlock displays
    - interlock mode for device correct

- LOOP TRACKING INTERLOCKS
    - loop track conditions match those in the detail design document
    - conditions included on an interlock display
    - condition verbiage and color expression
    - track value
    - manual override

- ALARM CONDITIONS
    - alarms match those activated in the detail design document
    - confirm the alarm functionality

- HISTORICAL DATA
    - trend collection and display is functioning per the design documents

- OPERATOR INTERFACE
    - faceplate and detail displays data as defined in the detail design document
    - dynamos are confirmed as displaying the correct data

- CASCADE LOOP CONTROL
    - master loop
    - slave loop
    - master mode to auto; slave mode to remote set point
    - slave mode to remote set point; master mode to auto

- slave mode not remote set point; master to output track slave process variable
- master to manual; slave to auto (one-shot)
- ADVANCED LOOP CONTROL
    - verify calculations and logic

Section 4 – Discrete Element Control (Motor Start/Stops, On-Off Devices)

4.1 Purpose – general statement to describe the Discrete Element Control functionality.

4.2 Test Equipment – list any test equipment required (e.g., signal generators, multimeters, jumpers, switches).

4.3 Procedures – a form-based procedure or checklist is helpful that can be documented per the Section 2 methods:

Discrete Control Element Testing

- NORMAL OPERATION
    - access field
    - mode flag
    - fail flag
    - interlock flag
    - bypass flag
    - disable flag
    - device color expression
- PROCESS CONDITIONS
    - all necessary variables configured
    - deadband functionality
    - verbiage on interlock/help displays
    - color expression on interlock/help displays
- INTERLOCKS
    - interlock conditions match those in the detail design document
    - tag color expressed properly on interlock displays
    - interlock action for device correct
- MONITOR FORCE CONTROL
    - monitor force conditions match those in the detail design document

- action taken
- monitor force type
- manual override
- enable/disable functionality

- ALARM CONDITIONS
  - alarms match those activated in the detail design document
  - confirm the alarm functionality

- HISTORICAL DATA
  - trend collection and display is functioning per the design documents

- OPERATOR INTERFACE
  - faceplate and detail displays data as defined in the detail design document
  - dynamos are confirmed as displaying the correct data

Control Strategy Testing

- DIFFERENTIAL GAP CONTROL
  - high-level, low-level, and deadband variables
  - high- and low-level control
  - manual override

- LOOP TRACKING
  - loop track conditions match those in the detail design document
  - conditions included on an interlock display
  - condition verbiage and color expression
  - track value
  - manual override

- MONITOR FORCE CONTROL
  - monitor force conditions match those in the detail design document
  - action taken
  - monitor force type
  - manual override
  - enable/disable functionality

- PREDETERMINING TOTALIZERS
  - reset functionality
  - transfer path opened
  - low flow cutoff

Loop Checking

- low flow cutoff functionality
- flow element
- totalizes in proper units
- hold functionality
- preact
- preact functionality
- transfer path closed
- complete flag

- CASCADE LOOP CONTROL
    - master loop
    - slave loop
    - master mode to auto; slave mode to remote set point (CASCADE)
    - slave mode to remote set point (CASCADE); master mode to auto
    - slave mode not remote set point (CASCADE); master to output track slave process variable/set point
    - master to manual; slave to auto (one-shot) — depends on the process and the plant preference

- ADVANCED LOOP CONTROL
    - verify calculations and logic

Section 5 – Display Testing

5.1 Purpose – general statement to describe the operator displays to be tested and the overall operator interface philosophy.

5.2 Procedures – a form-based procedure or checklist is helpful that can be documented per the Section 2 methods:

Display Testing

- displays
- all display elements that have not already been tested
- display hierarchy
- layout
- nonterminating lines traced to next/previous display via access fields

Section 6 – Test Results

6.1 Summary – per Section 2 methods, the checklists and discrepancies are listed here along with any corrective action notes.

6.2 Sign-off Sheet

Documenting the Results - Checklists and Discrepancy Forms

Technicians can develop checklists to keep track of the items in the preceding list during the testing. Forms for noting discrepancies should also be maintained.

2.4 PROCESS SIMULATION

As a further note on simulation, all control logic of any significant complexity should be tested by utilizing simulations. Simulations will give plant personnel and the implementation team the means to verify the accuracy of the logic that is spelled out in the process descriptions and to verify that the control logic performs the actions as designed. Plant personnel have said that simulation testing is very valuable, not only for the project FAT/SAT cycle by providing the means for more complete and accurate testing and compression of project schedules but also for on-going proof testing of new and advanced strategies. By completing these checks prior to shipment to the site, the chances for project success are greatly enhanced by providing a non-stressful environment to correct problems before plant production/start-up schedules may be affected as well as allow training for plant personnel on the new control system.

Software packages are available that allow the control system's I/O to be replaced by a PC and software that emulates I/O and the process (see Figure 2-2). Technicians can utilize a "tieback" software program to simulate the action of each discrete element and to simulate the action of simple loops (i.e., flow, pressure). Mass balance models can also be written for the individual process units in which the control logic performs ingredient additions and product transfers from a vessel or vessels. The simulation software can also be used to model complex chemical reactions and process thermodynamics if technicians cannot verify the proper operation of process logic without them.

Loop Checking

FIGURE 2-2
Process Simulation Architecture

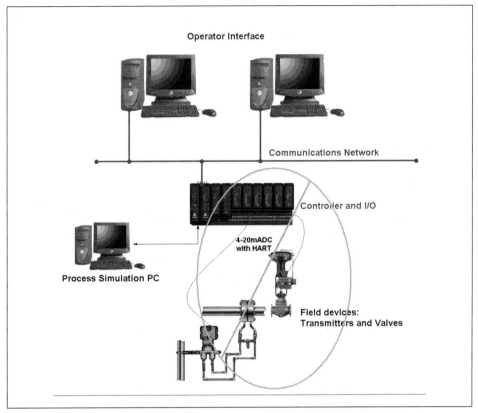

Technicians can also wire hardware using analog signal generators, multimeters, and switches/lights into the control system I/O to simulate process inputs. However, this method usually limits the complexity of the tiebacks and the simulation.

Technology Improvements

The impact of "smart" instruments is probably the largest on the site acceptance test (SAT). The factory acceptance test has the advantage of the PC-based simulation systems or signal generators for "easily" setting signal levels to verify the more complex interlock and interactive control strategies during the testing. Since the SAT occurs after the system is installed at the plant and field wiring is connected, the job of testing the more complex interlock and interactive strategies is very difficult due to the geographic distances between devices that need to be connected to/ manipulated/disconnected for testing.

As we'll see in Chapter 3 on start-up loop checking, remotely located devices can communicate with, and set, the smart devices to a test mode that allows test signals to be generated to verify operation. In fact, some smart instrument "asset management" packages have capabilities to automate and document this process for the many instruments involved in a complex strategy – see Figure 2-3.

FIGURE 2-3
Example SAT Loop Check

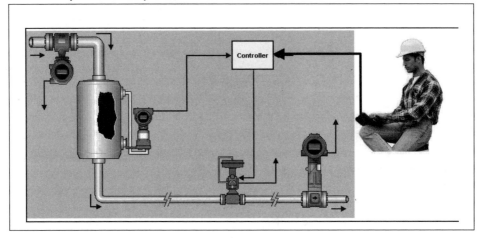

2.5 PROCESS CONTROL EXAMPLE

We will use the following example of boiler drum level, feedwater, and steam flow (three-element control) from Chapter 1 to illustrate our loop check at the FAT.

Documentation

The required documentation includes:
- drum level, steam, and feedwater P&ID / EFD
- SAMA drawing for drum level control
- detailed design specification for drum level control strategy
- loop sheets / wiring drawings
- reference diagrams / write-ups (see Chapter 1)
- hardware bill of material
- cabinet layout and power/grounding drawing
- system architecture drawing

Test Plan

The hardware and loop check test plans as outlined in this chapter will be used. The following specifics will be added to the plan checklists.

Loop Checking

The drum level control strategy test plan will verify:
- inputs and outputs
- feedwater flow loop action
- drum level control loop action
- steam flow action
- mode indications and alarms

Test Equipment. The technician will use a simulator system to manually force all inputs and monitor outputs. He or she will also perform a first-principles simulation of the steam drum, steam flow, and feedwater flow systems. The technician will use the control system operator interface screens to check indications and control actions.

Test Methods. The following four test methods will be used:

1. Inputs and outputs: Using the simulation program, the technician should force inputs to 0%, 25%, 50%, 75%, and 100% of span and verify the operator interface indication. The technician should note alarm points and also verify that they indicate on the operator interface. Outputs should be manipulated by placing loops in manual and setting them at 0%, 25%, 50%, 75%, and 100%. The simulation program can be used to verify that the proper output channel is changing to the desired value. The technician should indicate on a checklist when items are completed, adding any applicable comments or noting any discrepancies.

2. Feedwater control: With the loop in manual, open the feedwater valve and verify that the simulation increases the feedwater flow. The technician should place the loop in automatic and make set point changes, verify flow controls to set point, and then place the loop in cascade.

3. Drum level control: With the loop in manual, the technician should change the output and verify that the feedwater loop set point follows and controls. Next, he or she should place the loop in automatic and make set point changes, then verify level controls to set point.

4. Steam flow feedforward: The technician should make changes in steam flow and verify that the feedwater flow set point flows the steam flow and that drum level maintains the set point.

Checklists

The following worksheets are examples of this testing methodology. In particular, Tables 2-2 through 2-4 are examples of the test flowchart check areas.

TABLE 2-1.
General Test Checklist

CHECKLIST TASK	COMMENTS/DISCREPANCIES	CUSTOMER	VENDOR
1. Design Review and Documentation			
1.1 P&I Drawings 1.2 Design Specs 1.3 SAMAs 1.4 Loop Sheets			
2. I/O Testing			
2.1 I/O Configuration 2.2 Diagnostics Check 2.3 Test Signal I/O and Communications			
3. Test Plan			
3.1 Analog Elements			
3.2 Discrete Elements			
3.3 Conditions/Interlocks			
3.4 Discrete Element Control			
3.5 Regulatory Loop			
3.6 Fixed Sequences			
3.7 Advanced Control			
4. Operator Displays			
4.1 Layout and Design			
4.1 Layout and Design			
5. Simulation			
5.1 Design & Complexity			

Loop Checking   43

FIGURE 2-4
Example Hardware Checklist – see Ref. 5 (Courtesy of Emerson)

BIOTECH PHARMACEUTICALS CORP.
MANUFACTURING CONTROL SYSTEM
REV. 0

ENCLOSURE TAG:
INSPECTED BY:
DATE:

REFERENCE DWGS:

EMERSON
Process Management

PERFORMANCE SOLUTIONS
LIFE SCIENCES INDUSTRY CENTER
PC&E DIVISION

PRE-ENERGIZATION FACTORY INSPECTION CHECKLIST

ITEM	DESCRIPTION	INITIALS	DATE	COMMENTS
1	RECEIVE COPIES OF FABCON IN-PROCESS INSPECTION CHECKLIST (QF 190-2), FINAL INSPECTION CHECKLIST (QF 190-3) AND APPLICABLE PNEUMATIC TUBING LEAK TEST REPORTS (QF 191-1) AND CONFIRM APPROPRIATE ITEMS ARE INITIALED, SIGNED AND DATED.			
2	VERIFY ALL HARDWARE IS INSTALLED IN ACCORDANCE WITH THE IFC DRAWINGS.			
3	VERIFY THE POWER AND CONTROL CIRCUITS ARE INSTALLED IN ACCORDANCE WITH THE IFC DRAWINGS.			
4	VERIFY NAMEPLATES HAVE BEEN ENGRAVED IN ACCORDANCE WITH THE IFC NAMEPLATE SCHEDULE.			
5	VERIFY THE ENCLOSURE CONTAINS THE UL LABEL AND ALL UL SAFETY STICKERS HAVE BEEN INSTALLED.			
6	VERIFY ALL WIRES ARE PROPERLY ROUTED AND LABELED IN ACCORDANCE WITH THE IFC DRAWINGS.			
7	VERIFY ALL FUSE AND CIRCUIT BREAKER SIZES ARE IN ACCORDANCE WITH THE IFC DRAWINGS.			
8	VERIFY ALL FOUNDATION FIELDBUS AND DEVICENET BULKHEAD FITTINGS HAVE BEEN INSTALLED WITH THE CORRECT GENDER ON THE EXTERIOR OF ENCLOSURE IN ACCORDANCE WITH THE IFC DRAWINGS.			
9	VERIFY FANS AND EXHAUST FILTERS HAVE FILTER ELEMENTS INSTALLED.			
10	VERIFY TERMINAL BLOCKS ARE MARKED IN ACCORDANCE WITH THE IFC DRAWINGS.			
11	VERIFY THE CORRECT QUANTITY AND LOCATION OF DELTAV CARDS ARE INSTALLED WITHIN THE DELTAV CONTROLLER CABINET ACCORDING TO EMERSON DIRECTION.			

Page 1 of 1

FIGURE 2-5
Example Loop Function Test Form

No.	Item Description	Expected Result	Actual Result	Pass/Fail
Non-Remote Modes (MAN &AUTO)				
1.	Ensure that the Mode is set to 'MAN'. Set OUT parameter either by numeric entry or by using the Output slider.	A) The mode can be switched to MAN using the button on the faceplate. Actual mode is MAN.	A) Target Mode is _____ Actual Mode is _____	A)
		B) Output is set.	B) Output is _____	B)
		C) The current output is indicated by a bargraph and a numeric display.		C)
2.	Switch the Mode to AUTO using the button. Enable PID1/Simulate, and give some value to PV.	A) The mode can be switched to AUTO using the button on the faceplate. Actual mode is AUTO	A) Target Mode is _____ Actual Mode is _____	A)
		B) PV is indicated on the PV Numeric Indicator and Bar graph	B)	B)
		C) Mode can be changed to AUTO	C)	C)

TABLE 2-2.
Analog Element Checklist

Type	Tag	Description	Device	Address	Display	EU	EU100	EU0	Access	PV field	Bargraph	EU	Alarm
ANALOG IN	FI-101	Steam Flow	CNTRL-1	1-1	Boiler 1	KPPH	75	0	ok	ok	n/a	ok	ok
ANALOG IN	PI-102	Drum Pressure	CNTRL-1	1-2	Boiler 1	PSIG	500	0	ok	ok	n/a	ok	ok
ANALOG IN	TI-103	Steam Temp	CNTRL-1	1-3	Boiler 1	DEG F	750	0	ok	ok	n/a	ok	ok

Notes:
- **Access**: At the operator workplace, call up the Display and access the tag using the mouse noting proper indication that tag is selected
- **PV field**: access the Simulation system at the address of the input point, change the value to a couple
- **Bargraph**: while accessing/changing PV field, verify that any associated bargraphs also move correctly
- **EU**: Verify correct Engineering Units (EU) are displayed
- **Alarm**: verify any alarms are correctly displayed

TABLE 2-3.
Discrete Element Checklist

Type	Tag	Description	Device	Address	Display	Off Word	On Word	Access	PV field	Alarm
DISCRETE IN	HS-101	DRUM XMTR SEL	CNTRL-1	2-1	Boiler 1	Xmtr A	Xmtr B	ok	ok	n/a
DISCRETE OUT	LSL-102	DRUM XMTR LOW	CNTRL-1	3-1	Boiler 1	NORMAL	LOW	ok	ok	n/a
DISCRETE OUT	LSH-103	DRUM XMTR HIGH	CNTRL-1	3-2	Boiler 1	NORMAL	HIGH	ok	ok	n/a

Notes:
- **Access**: At the operator workplace, call up the Display and access the tag using the mouse noting proper indication that tag is selec
- **PV field**: access the Simulation system at the address of the input point, toggle the input channel and verify display indication
- **Alarm**: verify (if applicable) alarms are correctly displayed on toggle

TABLE 2-4.
Basic Regulatory Loop Checklist

Type	Tag	Description	Device	Address	Display	EU	EU100	EU0	Access	PV field	SP field	OUT field	Mode field	Bargraph	EU	Alarm
REG LOOP	LIC-101	Drum Level	CNTRL-1	1-1	Boiler 1	INCHES	15	-15	ok	ok	ok	ok	ok	n/a	ok	ok
REG LOOP	FIC-102	Feedwater Flow	CNTRL-1	1-2	Boiler 1	GPM	250	0	ok	ok	ok	ok	ok	n/a	ok	ok

Notes:
- **Access**: At the operator workplace, call up the Display and access the tag using the mouse noting proper indication that tag is selected
- **PV field**: access the Simulation system at the address of the input point, change the Process Variable value to a couple values (e.g. 25%, 75%) and verify display indication
- **SP field**: access the tag, change the setpoint value to a couple values (e.g. 25%, 75%) and verify display indication and that simulation moves output and PV correctly
- **OUT field**: access the Simulation system at the address of the output point, change the output value to a couple values (e.g. 25%, 75%) and verify display indication
- **Mode field**: access the tag, change the mode to valid modes (auto, man, cas) and verify display indication
- **Bargraph**: while accessing/changing PV field, verify that any associated bargraphs also move correctly
- **EU**: Verify correct Engineering Units (EU) are displayed
- **Alarm**: verify any alarms are correctly displayed

REFERENCES

1. ISA-5.1-1984 (R1992), *Instrumentation Symbols and Identification*. ISA – The Instrumentation, Systems, and Automation Society, 1992.

2. ISA-5.4-1991, *Instrument Loop Diagrams*. ISA - The Instrumentation, Systems, and Automation Society, 1991.

3. ISA-TR20.00.01-2001, *Specification Forms for Process Measurement and Control Instruments Part 1: General Considerations - Updated with 20 New Specification Forms in 2004*. ISA - The Instrumentation, Systems, and Automation Society, 2001.

4. Cole, Wayne E. "World-class engineering delivers higher project quality," *TAPPI Journal* (Vol. 76, No.2), 1993.

5. Kramer, Greg. "Reduce Startup Time." Emerson Process Management, 2004.

QUIZ

1. What are some of the benefits of factory acceptance testing (FAT)?

2. What are the basic elements of a FAT?

3. How do PC-based Process Simulation Packages fit in a FAT plan and why are they beneficial?

4. How can smart instruments help in the acceptance testing?

3
START-UP

You have accepted the control system at the vendor's factory, and it has been delivered and installed. Appropriate power and grounding checks have been performed, and the vendor has signed off on it as ready for operation. It is assumed that field measurement and control devices (or transmitters and valves for our example) have been received, their calibration checked and documented, and that they have been installed in the process piping and wired to the control system's I/O. As the start-up date approaches, technicians perform loop checks to verify that the control system will perform as expected with the actual inputs and outputs connected. Some plants' implementation teams require that a site acceptance test be run to again verify all displays and control strategies using live signals from the field devices. Since this test resembles the factory acceptance test (FAT) described in the previous chapter, we will not describe again the complete FAT for verifying control strategy.

Since the process is not yet running, at this phase the loop check verifies the integrity of the wiring and the correctness of the control system's communications and display. Several scenarios for loop checking may be possible at start-up, depending on whether the system is a new installation, an expansion, or an upgrade of an existing system. Our review in this chapter is based on a new installation, which is probably the most comprehensive.

Note that we are discussing just the control loop here. Additional testing should be done for motor start/stops, interlocks, indicate-only type controls, and the like.

3.1 DOCUMENTATION

The technician should have a set of loop sheets that detail the field wiring input and output locations in the control system (see Ref. 1, 2, and 3). The P&IDs are also a helpful reference as is any vendor information on the field devices and control system I/O. It is desirable to gain a listing of the operator displays showing the location of the loops so technicians can verify the displays along with the control strategy drawings. This will enable them to know what is expected at the operator level.

3.2 LOOP CHECK PLAN

A loop check test plan will ensure that technicians get the most out of the time they spend testing the system. Typically, several people may be involved since one or more people are working the field devices and another on the control system operator interface. As noted previously, the plan should include the list of the loop I/O to be tested, the test actions, and the procedure for documenting the check (usually, the person at the control system operator interface signs off that the loop is complete). Everyone on the checkout team should understand his or her role. The plan should take the physical location of the installed instruments into account so an efficient route can be determined for the field personnel. Radios for communicating between team members are essential.

Note that the plan should clearly specify any signal characterization required. For example, "head" type flowmeters, which create a measurable differential pressure or "pressure head" in the fluid, generate a signal that has a square root relationship between flow rate and differential pressure (Ref. 4). Typical "head" meters include orifice plates, venturi tubes, weirs, and so on. Thus, the square root of the signal must be extracted either in the transmitter before the signal is sent or in the control system. The control system team member should be aware of these applications and verify the proper signal is received at the system. Plant standards can help here; for example, all square roots shall be applied in the control system. Many a troubleshooter, however, has seen problems in production when no square root is taken or even where the signal is square rooted twice—once in the transmitter and again in the control system.

3.3 CHECKING THE LOOP

We will describe the loop check based on the type of field instruments in use. The traditional instruments are 4-to-20-mA DC-based devices. Smart Instruments use a digital communications protocol that is transmitted on the individual 4–20 mA wires (e.g., HART®) or bus-based communications with multiple instruments on a communication cable (e.g., FOUNDATION™ Fieldbus).

Traditional Instruments (Non-smart)

Starting with the process measurement, the loop check usually begins with the "field" team member locating the transmitter in the piping. He or she should verify the location, tag, and installation details (such as ensuring that the impulse piping, grounding, wiring polarity, and manifold valves are correctly set). He or she should also note any comments on a checkout form.

At the transmitter end, the technician should connect a 4–20 mA source device (several manufacturers, such as Fluke and Altek, offer this device) to the two wires going back to the control system. Be sure to verify that the screw terminals have been tightened after checkout! Experience has shown that loose terminals cause more problems than anything else. The checkout team member at the control system operator interface is then informed that the transmitter is ready to be checked.

(Note: some plants prefer to perform the loop check in two steps. The first is to check the wiring from the field device to the control system I/O terminations in the rack room by lifting the wires at both the transmitter and rack room ends, shorting together at the transmitter, measuring the loop impedance at the rack room terminations, and also verifying that there are no shorts to ground. The display check from the I/O termination to the operator display is then performed at another time.)

The technician at the control system should have located the display where the transmitter variable is displayed (there may be several displays). Once the field person is ready, he or she should call for 0-25-50-75 and 100% steps of the transmitter output ("shooting a signal") from the field team member. There should be a pause at each step to confirm that the display is showing the correct value. The technician should be aware (and note on the checkout form) any input signal characterization during this check, such as square root extraction, pressure-temperature compensation, and the like. Such characterization may cause the operator interface reading to differ from the signal being introduced at the field device. Any alarms on the signal are also verified while the signal is stepped through the range. If satisfied that the transmitter and control system are performing as expected, the control system person checks this device off as verified and informs the field person to reconnect the wires and move on to the associated valve.

Note that some implementation teams prefer to check the transmitter output by actually simulating the process signal rather than just verifying wiring as discussed above. For example, a decade box could be connected to the RTD temperature transmitter input wiring or a pressure source to the impulse piping of the pressure/differential pressure device. This box would also be a good way to check transmitter calibration, although it should not be a substitute for good shop calibrations using certified source and measurement equipment. A veteran of a recent large job informed the author that tens of thousands of dollars in time and equipment had been spent only to find one bad transmitter and two with an incorrect range.

Technicians check the control valve in much the same way as the transmitter: at 0-25-50-75-100% steps. Again, the field team member should note any installation discrepancies since studies have shown that valve installation and setup issues can result in start-up delays and on-going performance problems (Ref. 5). After checking for instrument tubing crimps, broken pressure gauges, clean vent connections, etc., the technician should notify the control system team member that the valve is ready for the check.

The person at the control system should have located the valve on the display, accessed the loop output, and, after notifying the field person, made the 0% to100% step changes in the output (implied valve position). During each step pause, the field person should verify the valve position on the valve stem/shaft indication. Note that this step test is not an adequate indication of valve performance, which is discussed in Chapter 4. After establishing that the operation and display are satisfactory, the control system person checks this loop off the list.

This procedure is then repeated for each loop in the control system. The completion of this check now gives confidence to the team members that when the start-up occurs, this loop can control in manual. Loop commissioning/tuning represents a separate discussion.

Technicians should check critical interlocks and monitor/force (e.g., low tank level stops the agitator) type controls using a similar procedure to trace the input all the way through to the shutdown of the output device.

Smart Instruments

Smart field devices make the loop checkout more efficient. The procedure is the same: step changes in inputs and outputs are verified at the control system's operator interface. Theoretically, only one person is needed at the control system to "talk" to the smart field devices and "shoot the step signals" by using one of several types of communication tools available from the instrument and control system vendors. For example, the transmitter can be placed in "loop test" mode and the output to the control system set at the % steps. Likewise, the valve key parameters such as stem position and air pressures can be monitored while the control system outputs to the valve. These actual readings are usually more accurate than the traditional check's visual methods. In fact, just by communicating with the field device, the system can verify the wiring and perform tag number checks. It can also note any diagnostics problems with the field device and fix them prior to start-up.

Using smart instruments also reduces documentation time since the software packages typically have an "audit trail" that records the actions to each device, replacing the cumbersome manual entry of notes into a filing system by a team member. A technician uploads field device configuration parameters into the software database so they can be verified against spec sheets. This is a real time saver when you need to verify ranges and tag names while also looking for possible problems—such as incorrectly entered high damping values, which affect the control loop's performance.

In the actual field situation for a new installation most implementation teams make the trip to the field devices in order to verify that the right instrument has been mounted and that installation details are correct. However, this could be done before the loop check. Some implementation teams have "lifted" the wire at the device and verified at the control system that the transmitter value shows bad while valves are stroked from the control system with the field person watching the valve's movement. As mentioned earlier, technicians should verify that the screw terminals are tightened and good wire connections are made!

3.4 PROCESS CONTROL EXAMPLE

The following example of boiler drum level, feedwater, and steam flow (three-element control) is used to show a loop check during start-up (see Figure 3-1).

Documentation

FIGURE 3-1.
P&ID

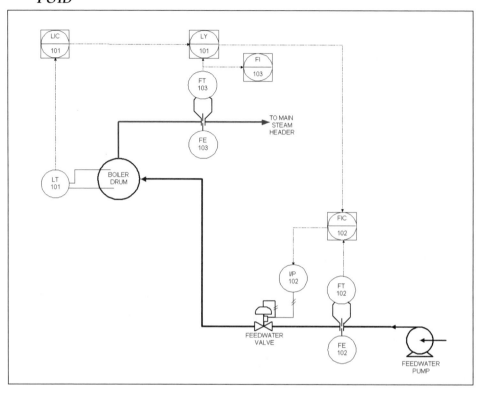

Test Plan

Traditional Instruments (non-smart)

Verify the feedwater loop, drum level, and steam flow inputs.

Feedwater Loop

Field team member: Proceed to the ground floor location of the feedwater flow transmitter, Tag FT-XXX, and connect a 4–20 mA source to the wiring at the transmitter. Radio the control system team member that you are ready to "shoot the transmitter" signal and upon his or her OK, step the signal from 4-8-12-16 to 20 mA.

Control system team member: Locate the feedwater loop on Display #YY and access the loop faceplate on the display. Remember to research the signal characterization for this "head" type flowmeter to know that the signal should be square rooted in the control system (per this plant's standards) and will be different than what the field member is calling out.

Loop Checking 53

For example, a 25%/8-mA signal from the transmitter will read 50%/50 kpph when communicated to the operator faceplate. Record the values from the display and faceplate on the loop check form (they should be 0-25-50-75-100 kpph), and note any comments made by the field team member on installation or other discrepancies.

Field team member: Proceed to the feedwater valve, which is located downstream of the transmitter. Make sure you can easily watch the control valve's stem position indicator, and radio the control system team member that you are ready for the check.

Control system team member: Locate the feedwater loop on Display #YY and access the loop faceplate on the display. Place the loop in manual and make output changes of 0-25-50-75 and 100%. Be sure to pause with each step and ask for verification from the field team member. Note each step on the loop check form along with any comments made by the field team member on installation or other discrepancies.

Drum Level Transmitter

Field team member: Proceed to the top floor location of the drum level transmitter (near the boiler drum), Tag LT-XXY, and connect a 4–20 mA source to the wiring at the transmitter. Radio the control system team member that you are ready to "shoot the transmitter" signal and upon his or her OK, step the signal from 4-8-12-16 to 20 mA.

Control System team member: Locate the drum level loop on Display #YY and access the loop faceplate on the display. Record the values from the display and faceplate on the loop check form (they should be –15.0 to –7.5 to 0.0 to 7.5 to 15.0 inches) and note any comments made by the field team member on installation or other discrepancies.

Steam Flow Transmitter

Field team member: Proceed to the top floor location of the steam flow transmitter (downstream of the drum steam line to the process), Tag LT-XXZ, and connect a 4–20 mA source to the wiring at the transmitter. Radio the control system team member that you are ready to "shoot the transmitter" signal and upon his or her OK, step the signal from 4-8-12-16 to 20 mA.

Control system team member: Locate the steam flow on Display #YY and access the indicator faceplate on the display. Record the values from the display and faceplate on the loop check form (they should be 0-25-50-

75-100 kpph), and note any comments by the field team member on installation or other discrepancies.

Smart Instruments

The approach to the test plan for smart instruments is similar to the traditional instrument method except that the checkout person is located at the control system's operator interface and has access to the smart instrument software package. (We'll assume that a field team member has visually checked each instrument for any installation problems.) Using the software package, a device scan verifies all instrument wiring (any missing instruments are then investigated for wiring problems) and tag numbers. The technician can check the field device configuration against the spec sheets by using the "compare" features of the software package's database. The transmitter and valve checks for 0%, 25%, 50%, 75%, and 100% of range can then be accomplished by using the smart instrument software while verifying the values on the control system displays. All these activities are documented in the smart instrument software package for future reference. Some plants have estimated time savings of around 30 minutes per loop using these smart instruments, not to mention possible safety benefits such as not having to work in the process area with nearly inaccessible instruments.

3.5 EXAMPLE FORMS

The following forms are examples that technicians could use during the loop check.

Loop Checking

CONTROL SYSTEM/INSTRUMENTATION CLOSED LOOP CHECKOUT

NOTE: One Instrument per Form

Date of Checkout:_____
Asset Number:_____
Instrument Number:_____
Instrumentation Location:_____
CONTROL SYSTEM Controller Number:_____
CONTROL SYSTEM File Number:_____
CONTROL SYSTEM Card/Slot Number:_____
CONTROL SYSTEM Channel Number:_____
CONTROL SYSTEM Cabinet Location:_____

TECH / DATE
_____ / _____ Verify DCS faceplate and graphics completion/operation
_____ / _____ Verify wires are terminated and labeled in DCS cabinet and field
_____ / _____ Verify that instrument is labeled
_____ / _____ Verify that air supply is installed
_____ / _____ Verify that conduit and flex are complete
_____ / _____ Verify that limit switch installed
_____ / _____ Verify that solenoid is installed
_____ / _____ Update electrical, loop, and/or EFD drawings; return to Engineering

TRANSMITTER
TECH / DATE
_____ / _____ Record faceplate indication at percentage value:
    0% (____ mA) = _____
    25% (____ mA) = _____
    50% (____ mA) = _____
    75% (____ mA) = _____
    100% (____ mA) = _____
Comments:_____

CONTROL VALVE
TECH / DATE
_____ / _____ Stroke valve (5 points) and verify positions
    0% (____ mA) = _____
    25% (____ mA) = _____
    50% (____ mA) = _____
    75% (____ mA) = _____
    100% (____ mA) = _____
Comments:_____

SOLENOID VALVE
TECH / DATE
_____ / _____ Open Pass____ Fail____
_____ / _____ Close Pass____ Fail____
Comments:_____

ENGINEER:_____
TECHNICIAN:_____
OPERATOR:_____
FCR CREW TECHNICIAN:_____

Loop Tag name		Description			Process Equipment #		
Transmitter				Control Valve			
Physical Address	A007	2-2-5		Physical Address	A007		3-2-2
Field Input:	*Record FP indication*	(√)		Loop Output:	*Verify Field Position*		(√)
0% (4 ma)				0% (4 ma)			
25% (8 ma)				25% (8 ma)			
50% (12 ma)				50% (12 ma)			
75% (16 ma)				75% (16 ma)			
100% (20 ma)				100% (20 ma)			
Date of Check:			ProcessControl	I/E Technician		Operations	
Comments:							

Loop Tag name		Description		Process Equipment #		
Transmitter			Comments			
Physical Address	A007	2-2-7				
Field Input:	*Record FP indication*	(√)				
0% (4 ma)						
25% (8 ma)						
50% (12 ma)						
75% (16 ma)						
100% (20 ma)						
Date of Check:			ProcessControl	I/E Technician		Operations

Loop Tag name		Description		Process Equipment #		
Process Switch			Comments			
Physical Address	A009	2-1-6				
Field Input:	*Record FP indication*	(√)				
De-energized						
Energized						
Date of Check:			ProcessControl	I/E Technician		Operations

Loop Tag name		Description			Process Equipment #		
Motor Start Output				Motor Status Input			
Physical Address	A009	2-5-6		Physical Address	A009		2-3-6
Issue Setpoint from Console:	*Verify Field Position*	(√)			*Verify FP Indication*		(√)
STOP				STOPPED			
START				RUNNING			
Date of Check:			ProcessControl	I/E Technician		Operations	
Comments:							

Loop Tag name		Description			Process Equipment #		
Solenoid Valve				Close Limit Switch Input			
Physical Address	A008	4-2-3		Physical Address	A008		2-3-4
				Open Limit Switch Input			
				Physical Address	A008		2-3-3
Issue Setpoint from Console:	*Verify Field Position*	(√)			*Verify FP Indication*		(√)
CLOSE				CLOSED			
OPEN				OPEN			
Date of Check:			ProcessControl	I/E Technician		Operations	
Comments:							

REFERENCES

1. ISA-5.1-1984 (R1992), *Instrumentation Symbols and Identification*. ISA – The Instrumentation, Systems, and Automation Society, 1992.

2. ISA-5.4-1991, *Instrument Loop Diagrams*. ISA - The Instrumentation, Systems, and Automation Society, 1991.

3. ISA-TR20.00.01-2001, *Specification Forms for Process Measurement and Control Instruments Part 1: General Considerations - Updated with 20 New Specification Forms in 2004*. ISA - The Instrumentation, Systems, and Automation Society, 2001.

4. Lipták, Béla. "Instrument Engineers Handbook", 1979.

5. Fitzgerald, Bill. "Control Valves for the Chemical Process Industries" McGraw-Hill, 1995.

QUIZ

1. What is the main objective of the loop check at start-up?

2. Briefly review the procedure for "shooting" the loop.

3. What key observation can the field team member make during loop checks?

4. What are some of the advantages of smart instrumentation?

4
PERFORMANCE BENCHMARKING

Once the control loop has been checked, started up, commissioned, and tuned, it is time to establish the baseline performance benchmark. Also, since control loop performance will degrade over time due to process changes, valve performance issues, and loop tuning changes, for example, it is beneficial to check the loop performance periodically. In addition, this "performance history" can be helpful when setting up predictive maintenance for your critical loops. If there is not a loop performance/predictive maintenance program in place at your plant, it's never too late to start by using the methods discussed here (and in the next chapter) to establish the current performance of the loop.

We'll assume this is a first-time "loop performance check" to include all areas of loop performance that should be checked. (Note: These procedures can also be used to troubleshoot problem loops.) Periodic testing to verify performance may just require a subset of the steps outlined below. The method we'll discuss to test performance involves the following:

- **Designing the Test** – Do your homework to understand control objectives, installation details, type and capability of the control equipment involved so test results will be meaningful. This will also help point out possible improvements.

- **Performing the Test** – Decide on a loop test methodology and proper tools that fit your plant's control system and your performance team's objectives. One such methodology we'll use includes (1) recording the as-found key parameter data to verify/document assumptions about the performance and system operation as well as apply analytical techniques to quantify the loop, and (2) performing open loop "bump" tests that provide a look into equipment performance as well as process dynamic information that can be used for loop tuning.

- **Analyzing and Reporting Test Results** - Turn test results into meaningful information about loop performance and make recommendations for improvement.

4.1 DESIGNING THE TEST

Documentation

As always, documentation is critical to understanding the loop-checking task. The technician should have access to loop sheets, piping & instrumentation diagrams (P&ID's), control strategy drawings, control system configuration printouts, transmitter calibration/range history, and vendor manuals so he or she can understand how the loop is intended to function. Prior to "designing" the loop performance check plan, these documents should be reviewed and understood.

Interview Key Personnel

Interviews with operations, engineering, and maintenance personnel are important to understand past history, what is expected, and any problems the loop is having. This information is the basis for the next step in planning. With a good knowledge of the history and operating problems, you'll have a better chance of establishing a plan that will produce a successful result.

For example, has the loop been operating in automatic? If not, why? What does the operator say the reasons are? How long has the loop not been in auto? Does Operations Supervision believe it is performing to their expectations? What is the expected performance of this loop? Most control systems have a process historian software package that can be used to help you evaluate and understand past performance. Engineering should also be consulted for their observations. Compare all inputs and resolve any plant differences on expectations.

Also, try to get financial information on the process such as cost per pound of product, cost of raw materials, cost of rework, etc. Then, arrive at an understanding with operations management on which of these costs are affected by the loop under consideration and what the savings would be after improvement. Typically, it is hard to pin an exact figure on improvement, so start with very conservative estimates such as 1% improvement on defective or rejected product, less raw material or increased production. These figures will come in handy, along with your recommendations on improving loop performance, to help management justify any needed expenditures.

Loop Check Planning

A loop performance check test plan will get the most out of time spent testing the loop. Typically, as the instrument technician conducting the test, you will need to be closely involved with the process engineer and operators/operations supervision. Make sure to involve them before the "loop check" testing starts.

The process control engineer can supply control strategy drawings and write-ups so you can understand the control objectives. Then, compare these with the actual control system configuration. It is not unusual for control implementers to have configured the strategies in ways that will affect how the loop will react. For example, it is important to know how the control system "affects" the measured variable before it reaches the control algorithm (PID), and if there are any changes to the output signal to the valve. For example, input filtering in the control system can mask problems along with process variable (PV) calculations that add non-linearity to the signal. Look for signal characterizations in the control system program that can affect the output signal to the valve.

From P&ID's and control strategy drawings, you'll want to understand how the loops interact in your application. This will help you decide the variables to analyze during your test. For example, if multiple loops are fed from a common header, then you might want to monitor all the loops to see if one is causing excessive header disturbance that can affect the other loops.

The maintenance input to planning is usually part of your job as instrument technician. You will need to identify I/O terminals, signal ranges, and location of field devices. Your knowledge of the control system's input and output circuitry will be needed to have the proper tools available for handling voltages required for the recording device. You will also probably be the best source to know any failure/replacement histories of field devices and any mechanical problems. Verify that the valve characteristic is adequate for the application and installation.

Your plan should now have a list of variables you want to monitor and test, along with appropriate ranges of the instruments. Based on your knowledge of the process, decide on the number and time duration of the "as-found" test runs. Knowing the number of open loop (controller in manual) bump tests, you'll have a feel for how long your initial testing will take.

There are many possible tools to help you gather and analyze data on the loop performance, including stand-alone Data Acquisition/Analysis/Tuning packages that "clip on" at the control system I/O, strip chart recorder printouts, hand sketches, stop-watch calculations, spreadsheets, and integral applications within the control system. With some of the more recent control system architectures, there are ways to "connect" to the process variables without "clipping on" at the I/O screw terminals. Standards such as OPC (OLE for Process Control) allow a fairly easy method of electronically sharing data directly between the control system and your data analysis package (more on this in the next chapter). Also, control system historians can collect data of interest and output to files that can be transferred (by floppy, CD, email) to your data analysis package or Excel spreadsheet.

Although this electronic method introduces some communication delays, signal filtering, and possible data compression, it is a lot easier than spending the day in a cabinet sorting out wires and looking for voltages. Depending on loop dynamics, this method can be adequate for a good percentage of our loops. With some critical loops, such as those around the paper machine headbox, getting to the raw, fast data at the measurement devices is still important. Remember your main job is to improve the loop's performance.

It is a worthwhile effort to evaluate the different analysis packages with your performance team in order to select one that is right for your plant.

Summarize and present all of the above information to the operators, since they will need to make control system operating parameter changes during your testing. Sometimes it can be a little disturbing to the operator when you ask if you can get into the wiring cabinets, put loops in manual, and stroke valves/output devices. Be sure they know you just intend to make very small changes (starting at 0.5%), they can put the loops back to auto at any time, and they won't even know you've been in the I/O room.

In summary, it's a good idea for the people involved with the loop check to know the check plan up front so they can prepare and make recommendations to help optimize the test time. Since you'll be manipulating the process, minimizing wasted effort is important.

4.2 PERFORMING THE TEST

When the scheduled time for loop performance checking arrives, verify with operations that the current production situation is as close to "normal" as possible. Most plants have unexpected problems or production constraints that could cloud any data you gather; it would be best to reschedule rather than make recommendations on questionable data.

The Walk-through

Having reviewed the documentation and drawings and set your plan, a "walk-through" of the loop process piping and installation is the next step. This should verify the impressions from your homework reviewing the documentation prior to visiting the plant site. Look for piping layout, installation issues and location of transmitter and valve. For example: What condition are the devices in? What is the valve supply air pressure? Any air leaks? Can you hear any cavitation? Are there straight runs of pipe before and after the measurement and valve? Are transmitter impulse lines correctly installed? Will the location of the devices introduce undesired deadtime? Record the transmitter and valve vendor model numbers and nameplate data on your Loop Inspection Data sheet (see typical sheet in Figure 4-1). Look for any sources of disturbances on your loop such as multiple users on a header. Digital cameras can record your installation (see Figure 4-2) and provide a means to enhance your report, send it to others for additional review, and provide a reminder after you've left the site.

Next, head for the control room to observe loop operation. What mode is the loop in? What is the output to the valve? What is the transmitter reading? If the measurement or valve position is near 0 or 100%, then hold off on the loop performance check until this condition is corrected. Record control system loop tuning parameters on your Loop Inspection Datasheet. Time and money constraints enter in, so you can't spend too much time hanging out with the operators, watching their actions, asking questions, and getting their opinions on problems and solutions. (Bringing a couple boxes of doughnuts in the control room seems to help this process.)

Note: It may be beneficial to perform the walk-through prior to the actual loop check test if time and costs permit. This would give you an opportunity to digest the installation information, along with the documentation, before designing the test plan.

FIGURE 4-1.
Example Loop Inspection Data Form

Loop Inspection Data

Customer:			
Plant/Mill:			
Area:			
Loop:	FIC-XXX	Field Engineer:	
Description:	Feedwater Flow	Date:	

Controller

Manufacturer:	Emerson DeltaV	Proportional:	0.40	Gain
Algorithm:	PID/Std/PI-error	Integral:	6.00	Sec/repeat
Scan Rate:	0.25 sec	Derivative:	0.00	Sec.
Engr. Units:	0-400 gpm	PV Filter:	0.00	Sec.
Notes: Mode= CAS Setpoint = 300.0 gpm		IVP= 55%		
Reverse; ARW <> = 0,100; IVP <> = 0,100				

Measurement

Manufacturer:	Rosemount	Model:	3051DP	Diff Pressure
Meas. Principle:	diaphragm	Range/Calibratn:	0-100	in wc
Protocol (Smart):	HART	Output:	4-20	mA
I/O Location:	Card 4 Ch03	Damping:	0.40	Sec
Installation Notes				

Final Control Element

Valve

Manufacturer:	Fisher	Model:	ET
Serial Number:		Style:	Globe/Sliding Stem
Travel:	3/4 in	Shutoff Class:	Class 4
Size:	2 1/2 inch	Characteristic:	Eq %

Actuator

Manufacturer:	Fisher	Model:	667-40
Serial Number:		Style:	Spring and Diaphragm
Air Pressure:	40 psi supply	Bench Set:	10-30 psi

Positioner and I/P

Manufacturer:	Fisher	Model:	DVC6010
Serial Number:		Input/Out Signal:	4-20 ma / 3-15 psig

I/P

Manufacturer:		Model:	
Serial Number:		Input Signal:	
I/O Location:			
Installation Notes			

FIGURE 4-2.
Document the Installation Through Digital Pictures and Field/Control Room Observations

Setting up for the Loop Check Test

For our discussions in this guide, we will test the loop using a high-speed data recording device and loop analysis software package (see Figure 4-3). Connecting to the transmitter and valve at the control system I/O termination blocks allows gathering the data (usually in a voltage format) for analysis without control system influences of filtering, reporting rates, or data compression that may cloud the data.

Set up your data-recording device to the control system input and output terminals for the valve and transmitter. Typically, a good workplace is difficult to find in the rack rooms where most I/O is located. An old wire spool can act as your workbench with an empty bucket from the trash can as your chair, but you might want to look into portable tables/work benches. (Note: The folding hand truck with table top in Figure 4-3, including power strip, storage for your tools, and large wheels to roll through/over obstacles to get to the rack room.) Plenty of extension cord with ground fault circuit interruption is a good idea, since the power outlets always seem to be across the room from the I/O cabinet where you are working. It is a good idea to put a small sign on the test area that includes names, phone numbers, plant contact, and time expected to have the test system in use. This lets plant personnel know testing is in progress and who to contact in case there is a need to perform maintenance or to troubleshoot in that area.

FIGURE 4-3.
Collecting Data with a Data-Recording Package

Connecting to the Process Signals

Depending on the age and design of the control system, plant documentation/wiring terminal identification, and plant wiring practices, this step can take some time. You can spend the good part of a day locating the input/output screw terminals for your loop, deciphering the control system I/O circuitry to determine the voltages available, and finding a suitable connection to "clip on" your data-recording device.

A common approach for control systems is to use a 250-ohm input resistor in the input circuitry to develop a 1–5 V signal for the A/D converter. By "clipping on" to this voltage, you can read the signal input of your data-recording equipment without interrupting the signal to the control system. Although sometimes the resistor may be easy to access, as shown in Figure 4-4, be sure to practice this method a few times before approaching a live process. While "clipping on/off" the control system, more than one process has been shut down due to carelessly loosening wires or yanking on non-soldered resistors.

Loop Checking

FIGURE 4-4.
Clip onto Voltage Signals at the Resistor

Other times the voltage is not as easy to access. The newer control systems are designing smaller I/O "footprints." Thus, the wire terminals are smaller and harder to get to once field wiring is installed. If test points are available at the I/O termination, such as in Figure 4-5, then some test leads help to clip onto the voltage.

FIGURE 4-5.
Clip onto Voltage Signal at Test Connections

If no "bare copper" is available, then "insulation piercing" type connectors are handy (see Figure 4-6).

FIGURE 4-6.
Clip-on with Insulation Piercing Connectors

Another alternative when no standard voltage is available, shown in Figure 4-7, uses a clamp-on, non-contact device that measures the current flow in the wire as the input to your data-recording device.

FIGURE 4-7.
Clip-on with Clamp-on Non-Contact Milliammeter

If smart (e.g., HART®) valves or transmitters are used, upload the device configurations into a smart field device software package or manually record the device setups using a hand-held communicator. (Note: Look for transmitter damping times. It is very easy to dial in a high damping value that will cover the true process dynamics. Same for the smart valve – check valve/actuator tuning parameters that could result in slow response.)

Also, the HART® transmitters and valves usually transmit other process measurement along with the main variable on the same wires that also can be used by your analysis with little extra effort to gather the data. For example, along with the control system 4–20-mA output to the HART® valve as the desired valve position, the same wire contains actual valve position, air pressures, and other diagnostic data. Coriolis meters can access flow density and temperature as well as the mass flow.

Other smart, all-digital protocols such as FOUNDATION™ Fieldbus and others present a particular problem in that there are no longer 4–20-mA current signals that get converted to voltages for measuring by your data-recording system. There are several vendors that include performance and analysis features in their digital bus-based system. One possible alternative would be to configure spare analog output channels in the control system to drive 4–20-mA current outputs representative of the signals of interest that could then be recorded.

Once you've completed the set-up, take time to verify your signals at the data-recording device with what the operator is seeing at the control system. Sometimes you've connected to the wrong terminal or the documentation is incorrect or the control system is "calculating" the display variable; in any case, you should feel confident about the data you are going to collect. This can prevent a lot of confusion later – there have been times when an alert process engineer presents system trends during your recommendation presentation that don't match your data values. Needless to say, this calls into question all of your data and any recommendations you've worked hard to get.

Running the As-found Loop "Time Series" Test

Now that you're ready to start collecting data, document your test runs with a lot of comments. (If your data-recording tool doesn't have notes fields, use note paper.) Include anything you think is important about the circumstances surrounding the test run – product names, production runs, time/date, operator comments/changes. All help when you are trying to make sense of the data in the analysis phase.

Record some time series data with the loop in the "as-found" condition (reference the examples below) to show how the variable is changing with time. The data, usually in engineering units, is plotted against time and also displays summary statistics for the variable at the bottom of the plot.

Collect some data while the controller is in manual mode. This gives a feel for the disturbances that the controller will see and provides a comparison with the loop in automatic. Ask the operator to make some set point changes if in automatic and also try to record during a disturbance to see how well the loop recovers.

You may want to make some overnight data runs to get longer collection periods in addition to the shorter runs. This will also help you to identify different period cycles that the process may be experiencing and give you an idea of any loop interactions that may be occurring. You now have a baseline of loop performance that you can compare against after any recommendations you make are implemented.

FIGURE 4-8.
Time Series Data

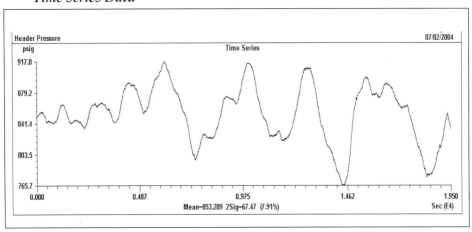

Figure 4-8 shows a typical time series graph of the data with statistics displayed. A cycle is noticeable in the data so that, by locating and eliminating the source, the performance could be improved.

Figure 4-9 shows a set point change to the process and the interaction between these boiler loops.

Running the Open Loop Check/"Bump" Test

Next, you're ready to run the open loop (loop in manual) bump tests. This will give you a couple of valuable pieces of information about the loop. First, it will allow you to calculate process dynamics such as gain, time constant, and deadtime that are important for loop tuning. The bumps will also give you a feel for the general condition of the control valve (or other final control element such as a damper, variable speed drive, etc.) and how well it's responding to the control system commands.

Set up your recording device to collect the process variable and the output from the control system (see Figure 4-10). Like the as-found data gathering test, add a lot of comments while your bump test runs. You'll

FIGURE 4-9.
Time Series Data

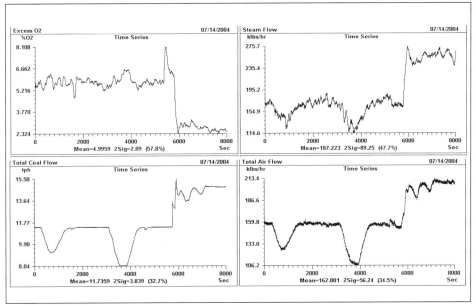

probably need to be sitting with the operator as you run this test since you will need to advise him when to go to manual (if approved by the operator) and when to make the bump changes in the output.

Here's how this might happen. After the operator places the loop to manual, ask him to change the output to the valve by 0.5% to 1% in either the up or down direction (whichever he is more comfortable with, based on where his process is). Make several steps in this direction, then have him do the same thing in the opposite direction. You should be looking at the transmitter variable to determine how many of these steps it takes for the process to respond. This gives a general idea of the valve's deadband and capability of being pushed with aggressive loop tuning.

A valve in good shape should respond to these small bumps. The time in between bumps will depend on the process response time. Let it settle out before making the next bump (remember, patience is a good thing here). Next, have the operator make larger (large is relative to the process and operator's comfort level) bumps that will be used in the process dynamic analysis. Several 2% to 5% (maybe larger for integrating processes such as levels) bumps are desirable to allow for a good average of the data. Turn the process back over to the operator and let things settle out before starting the next bump test.

The example in Figure 4-10 shows the loop output to the final control element in the upper half of the graph and the corresponding

FIGURE 4-10.
Typical Open Loop Bump Test Data

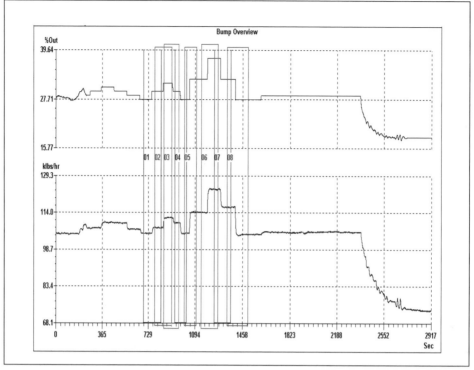

process flow response in the bottom half. (The analysis software allows you to "frame" each output bump/process response for analysis shown in the boxes labeled 01-08.) Notice the size and number of up/down bumps and also notice that the process is allowed to settle out before the next bump is made.

As previously mentioned, and with reference to the next chapter, some newer control systems are implementing these loop performance features as an integral part of the system (even automating the bump test), which greatly eases the time and effort of setup and testing.

4.3 ANALYZING AND REPORTING THE TEST RESULTS

Now that you have bump test and time series data, along with the walk-through information, you're ready to analyze the data, set a performance benchmark, and possibly make recommendations for improvement. Reporting (and storing for future comparison) your benchmark data and findings is an important consideration for working with your team when reviewing test results, making (and selling to management) any recommendations, and for sustaining the performance over time (more on this in the next chapter).

As-found Data Collection Analysis

Your data collection tool (or analysis software such as a spreadsheet) can help you pull out meaningful information from the raw data. There are a number of statistical techniques that have been used in process control for evaluating and setting benchmarks. For example, mean (average of the data), 2-Sigma (2 standard deviations which gives an idea of the spread of data around the mean), variability (2-Sigma divided by the mean) and variance, integrated absolute error (IAE), Harris Index, correlation, oscillation analysis, robustness, and financial priority/ assessments are just a few of the measurements that analysis tools can provide. You will need to work with your process improvement team to decide which are right for your plant.

Another valuable analysis tool that some packages supply is the "power spectrum" which breaks down the signal into the individual frequencies in the data so you can see which are causing the most variability. Process cycle/oscillation identification can be very helpful because the more defined cycles many times indicate loop tuning or valve sticking problems that can be identified and fixed. And, since a lot of loops are interacting with other loops in the unit operation, driving out the cycles will not only reduce the variability in the loop of interest but can improve the other loops that interact.

Other data analysis techniques, such as histograms (which show the statistical distribution of the data points over their range and around the mean or average value), and correlation analysis (which shows the degree to which the variability of a signal has a purely random characteristic or the degree to which 2 signals show a relationship with each other), can also help analyze the data to find variability sources.

Bump Test Analysis

As previously mentioned, analyzing the bump test will allow us to calculate the process dynamics (or how fast the process responds and how much), which is the critical data needed in tuning the loop. Also, the bump tests reveal how well the final control element is responding.

First, let's look at the process dynamics calculations. There are many resources that discuss loop tuning, process models, and the math involved (see Ref. 2 and Appendix A). We won't go into the detail here but will show an example of how the process dynamic information that needs to be determined for loop tuning might be calculated.

FIGURE 4-11.
Typical Data Analysis Tools

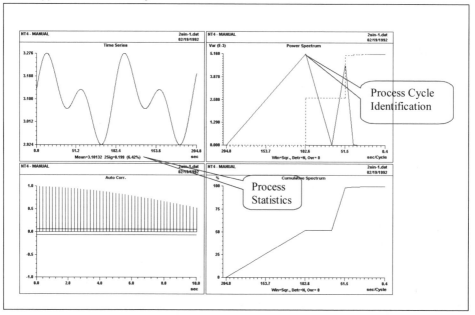

A good percentage of the loops that you'll be called upon to check and tune will fall under 2 basic types: 1st Order (or self-regulating) and integrating. 1st Order loops are typically flows and some pressures, while the integrating processes are levels (and sometimes pressures). Temperature loops, due to thermowell installations, are sometimes classified as 2nd Order. There will be other loops that you'll need to investigate with unusual dynamics – look to your analysis package vendor for further information on how this is handled.

The 1st Order loop dynamics are calculated from a graph similar to Figure 4-12 where the output is bumped (top graph - 1% in this case) and the process variable (PV) flow is monitored until it settles out (bottom graph flow changed about 2%). The calculations made for this type loop are:

- Process Gain (Kp)
- Time Constant (Tau, τ)
- Deadtime (Td)

Loop Checking

FIGURE 4-12.
1st Order Process Dynamics

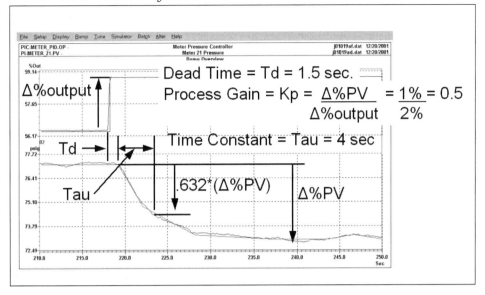

Similarly, the integrating loops (such as level) are evaluated as shown in Figure 4-13 and the calculations are

- Process Gain (Kp) = (final slope − initial slope)/change in % output
- Deadtime (Td)

FIGURE 4-13.
Integrating Process Dynamics

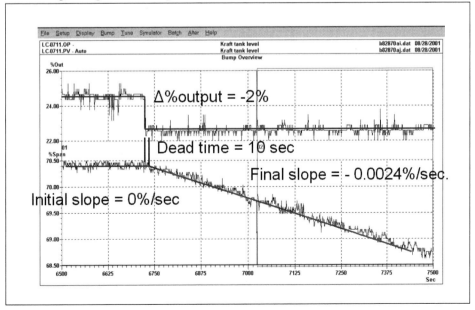

Loop Tuning

Now that the loop dynamic information is known, the loop tuning process can begin. As previously discussed in Chapter 1, one source showed that 30% of loop problems come from tuning issues. Appendix A on page 115 contains a very good discussion of loop tuning with background on loop basics, terminology, and several of the popular tuning methods. However, you'll probably find many opinions and preferences for tuning practices, and this guide is no different. One tuning method referred to as "Lambda" tuning has proved to be very effective in process control applications for chemical plants, power plants and pulp and paper mills where unit operations have a number of highly interactive loops. Lambda tuning provides a straightforward method of tuning that includes the flexibility to adjust the loop response to "de-couple" or minimize the loop interactions.

Lambda tuning is based on a model of the process to be tuned. Without going into much of the theory (see Ref. 2), by understanding the process to be controlled, the controller algorithm (PID, for example) and the desired response to set point changes/disturbances, you can then set tuning rules and be able to predict and analyze loop performance.

The key activity in the Lambda tuning method is the bump test and the resulting data previously shown. By making step changes (bumps) to the output of the controller while in manual, you can calculate the key parameters—Process Gain (K_p), Time Constant (τ), and Deadtime (T_d)—for tuning the loop. Lambda tuning gets its name from the Greek letter, λ, which is used in the tuning rules to set the closed loop time constant, or, in other words, to show how fast the loop will respond to set point changes/disturbances. The λ value you select is based on your knowledge of the desired loop response, interaction with other loops, confidence in your bump test results, and the ability of the loop devices to respond.

You can now apply the Lambda tuning rules to the bump test data. For example, 1st Order (self-regulating) loops:

1. Set reset (integral), T_r, to the open loop time constant, τ
2. Controller gain, $K_c = T_r / (K_p (\lambda + T_d))$
3. Rate (derivative), $T_D = 0.0$ sec

And for the integrating loops:

1. Set reset (integral), $T_r = 2\lambda + T_d$

2. Controller gain (proportional), $Kc = Tr / Kp (\lambda + Td)2$

3. Kp for integrating process = $\Delta y/\Delta t / \Delta u$ (where Δy = % change in the process over the Δt change in time for a Δu, step change in controller output or use the slope method shown in Figure 4-11).

4. Rate (derivative), TD = 0.0 sec.

Here are a couple of rules of thumb:

1. Start with a Lambda of 3 times the open loop time constant, τ. Confidence in your bump test parameters, low deadtime and high-performing control valves allow for faster Lambda values.

2. For interacting loops, or a cascade strategy, set the upper loop λ at 5–10 times slower than the lower loop.

By adjusting the Lambda value, the technician tuning the loops can apply his or her knowledge of the process (such as how the loop interacts with other loops or how well the valve responds to small changes to allow aggressive tuning) to get the desired results. The process example later in this chapter shows how this method can be applied.

Now let's talk about the additional information on the final control element (a control valve, damper, variable speed drive, etc.) that can be evaluated by the bump test. In Figure 4-14 bump test, notice in bump 07 that for an open command (bump up) the process variable did not respond until the second bump in the same direction. This is typically an indication of valve/damper deadband that makes it difficult for the loop to control closely to set point. Depending on your loop tuning strategy, deadband could prevent you from being as aggressive on the loop tuning as needed to maintain product specs. Other things you might see are unequal changes in process variable from equal bumps, different responses at different valve positions (e.g., 30% vs. 65% open) and varying deadtimes – some of the things you'll have to evaluate for your loop.

FIGURE 4-14.
Bump Test for Final Control Element Evaluation

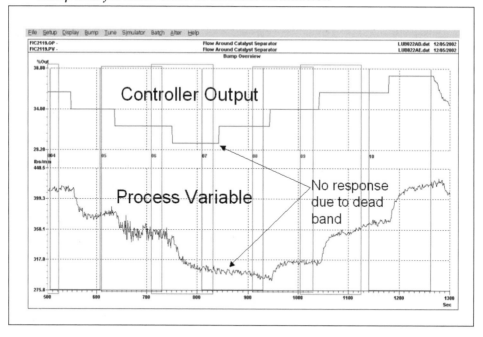

Reporting the Test Results

Some method of documenting and storing your tests and results should be decided upon. We have seen several approaches to this task.

- Customize your own with a word processor or database, cut and paste pictures from your analysis data and graphs, and manually transcribe results.

- Use commercially available Data-Recording and Analysis packages, which typically have a filing system for test and analysis results.

- Use Instrument Asset Management Packages, which also provide a good one-stop location to store the testing results since valve and transmitter diagnostics, setup data, and calibration data all relate to the analysis of the loop performance.

Some vendors provide more integrated loop testing and reporting in their products, so be sure to include this in your evaluations of your plant's loop checking program and tools.

Reporting your results to your team members, and to operations and management in a Summary Meeting, is also important to keep everyone aware of your progress and contributions. Frequently, you will want to implement the recommendations for the problems you find during the loop check. By adding financial data to the report such as costs, expected improvements, and return on investment to your test results, you'll be able to make a better case for gaining approval to proceed.

Fix/Tune and Re-run the Loop Check Test

Should there be recommendations for bringing the loop into performance specs, proceed with fixing any loop devices, process improvements, and/or tuning. Repeat the Loop Check testing to establish and document the results for variability, response to bumps and response in automatic. This is now your updated, latest benchmark for loop performance.

Next Step

Following the Summary Meeting, and having received Operations' approval on the performance benchmark, you are now in a position to move into the next phase of sustaining this performance over time. Depending on the severity of the service conditions, changing process conditions, and other factors, you will need to establish the methodology for sustaining the new performance. This is covered in the next chapter.

4.4 PROCESS CONTROL EXAMPLE

The following example of boiler drum level, feedwater, and steam flow (3-element control) is used to show an example loop check with performance benchmarking.

Designing the Test

Documentation

The P&ID is located and gives the technician an idea of the instrumentation and control loops involved (Figure 4-15). The technician will also pull the loop sheets to get an idea of where the wiring is terminated, relevant control system addresses, and ranges for the instruments. He or she will begin filling out the Loop Inspection data sheet (see typical data sheet on page 64) to keep track of the information.

FIGURE 4-15.
P&ID

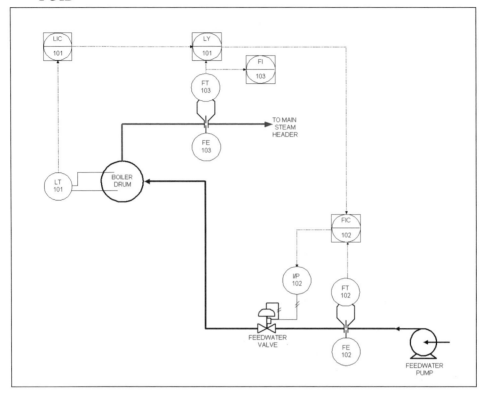

The technician then prints out a copy of the control system configuration so he or she can compare the latest with the Detail Design Spec from the factory acceptance test. The loop tuning parameters are recorded from the control system display and compared to the "as-left" documentation kept on file from the loop commissioning and start-up. It looks as though there have been some minor tuning changes made since start-up, but the overall configuration has been edited.

Interviews

Next, the technician visits the control room several times to visit with different shifts to get their perspective on what is expected. The operators indicate that they'd like to hold drum level within a couple inches during normal load swings and certainly avoid any trips due to high or low level. The technician mentions that he or she would like to do some testing to establish performance and set loop tuning by moving the feedwater valve in manual. The operator doesn't have any problems as long as he's watching the drum level and can stroke the feedwater valve. The powerhouse superintendent echoes these thoughts but places a dollar figure on the cost of unscheduled downtime by having to run the more expensive backup package boiler. The control engineer says the control strategy checked out okay in simulation at FAT. The engineer doesn't see any problems with operation, since the initial trends look okay, but the engineer would like to participate in the bump testing and loop tuning.

Planning the test

The variables that the technician needs to monitor are the drum level, feedwater flow, and steam flow along with the output to the feedwater valve. The actual feedwater valve position would be another good choice to ensure the valve is performing correctly during actual service. (Note: Since this is a smart valve, the loop check at start-up confirmed that actual position met desired position at 0, 25, 50, 75 and 100% travel points, but this does not necessarily indicate that the valve is responding in the 0.5% to 1% range.)

For this example, the technician is going to use a stand-alone Data-Recording/Analysis package that connects to the control system I/O. (As previously mentioned, there are several methods to collect and analyze your data.) From looking at the control system I/O circuitry, the data-recording "clip on" connectors should have easy access to the voltages the technician wants to monitor for the test.

Performing the Test

The Walk-through

As the technician walks around the boiler, the installation looks good. The feedwater transmitter and orifice plate are upstream of the valve; drum level and steam flow transmitters are mounted correctly, and all are smart HART® transmitters. The impulse lines meet our plant installation standards, and no leaks are noticed. The control valve is a sliding stem type with a spring and diaphragm actuator and smart digital valve controller for positioning.

It's a good idea to write down the valve nameplate data and/or serial number to have the vendor supply the assembly sheet with all pertinent valve data. With a typical boiler feedwater flow pump head and system loss curve, the technician will want to see an inherent valve characteristic of equal percentage. Mechanical integrity looks good, flow direction correct, no air/packing leaks are noticed, and air pressures at the air set match nameplate requirements. The technician records all this information on the Loop Inspection Data form.

Next, the technician visits the control room and talks with the operators about any problems, while noting set points and flow conditions. The drum level and feedwater loops are in Auto and Cascade, respectively, and plant production rates are at normal targets. Ask the operators to look at trends of the past several hours and call up the loop detail displays to record tuning parameters. Again, log the data on the Loop Inspection Data form.

Setting up for the Test

With the above complete and the list of test variables in hand, the technician sets up the Data-Recoding device in the rack room where the control system I/O terminals are located. He or she identifies the I/O terminals and clip onto the signals in the list – each time verifying the signal with the operator console to make sure he or she is on the right terminal. A HART® communicator is connected to each transmitter and valve to record and verify setup parameters. Start a test run to make sure the data is being recorded and looks correct. Then cancel without saving the data.

Running the As-found Time Series Test

With the drum level in automatic and feedwater flow in cascade mode and the first data recording run started, have the operator make set point changes to the drum level controller and monitor responses. Be sure to record while the steam flow load changes to see the effect on drum level variability. Run a test at fast sampling interval and then set up and run overnight at a slower sampling interval. This will give the technician data to analyze for fast and slow cycling of the data.

Running the As-found Bump Test

Perform bump tests on the feedwater flow loop and bring performance to specification through tuning or device repair. With the feedwater loop in cascade mode, drum level in manual, and steam flow feedforward disabled, bump the drum level loop output to determine the process dynamics of the boiler drum level. Tune the drum level loop to desired performance. With the drum level and feedwater loops in manual, record during a steam load change to see the subsequent effect on drum level to determine feedforward tuning parameters for gain (and lead-lag dynamic compensators, if needed).

If no major problems are found, and with loops tuned and in automatic/cascade mode, record time series data for the benchmark of drum level variability at various steam flow load changes.

Analyzing and Reporting the Test Results

As-found Data Collection Analysis

The time series data show the loops are performing in an acceptable manner (Figure 4-16). The Power Spectrum did not indicate any significant cycling with the loop in manual or auto, and the variables responded well to set point changes. A disturbance from steam load change was monitored, but the resultant drum level change was arrested in an acceptable amount of time/deviation and returned to set point.

FIGURE 4-16.
Time Series Data

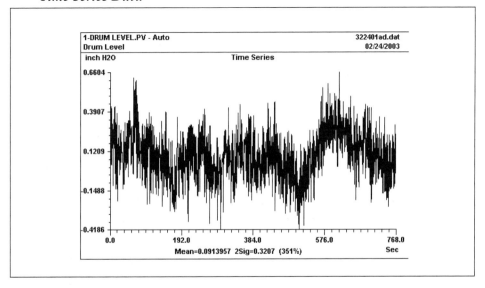

Bump Test Analysis

The bump testing allows the technician to analyze the condition of the final control elements (feedwater valve) and to determine process dynamics for loop tuning. The following graphs show some actual drum level/feedwater flow bumps with a fair-performing control valve (Figure 4-17) and a poor-performing control valve (Figure 4-18).

FIGURE 4-17.
Bump Test with a "Fair" Performing Control Valve

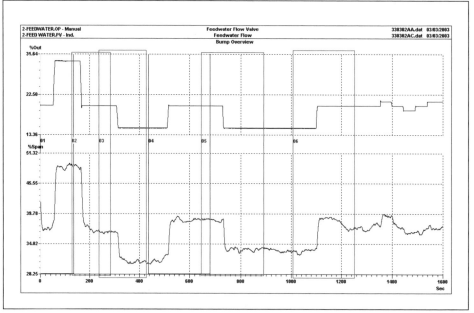

With a "fair" performing control valve, the feedwater flow follows the output even at the smaller bumps, although not as precisely as with a good/excellent performer. Note some of the gain differences in bumps 2-5. The level response is shown in the upper graph and shows the appropriate response to the increases and decreases in the feedwater valve.

The poor control valve doesn't respond to several of the output signal changes and thus causes a very sluggish level with significant deadtime. Process disturbances would be difficult to attenuate, and the loop would have to be de-tuned to prevent cycling. Of course, one of the recommendations at this point would be to repair or replace the feedwater control valve and re-test the loop.

The Bump Test also allows the technician to identify process dynamics to help in loop tuning. The drum level response can be considered an "integrating" process model while the feedwater flow is a self-regulating/1st Order response. Some example loop tuning displays from a Data-Recording and Analysis package are shown in Figures 4-19 and 4-20.

Loop Checking

FIGURE 4-18.
Bump Test with a "Poor" Performing Control Valve

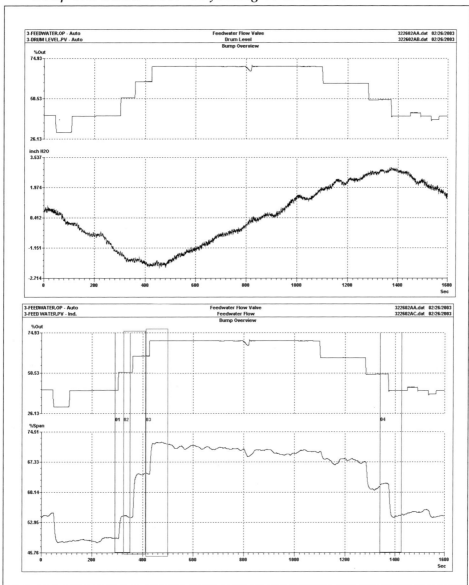

FIGURE 4-19.
Example Bump Test Analysis for a 1st Order Process

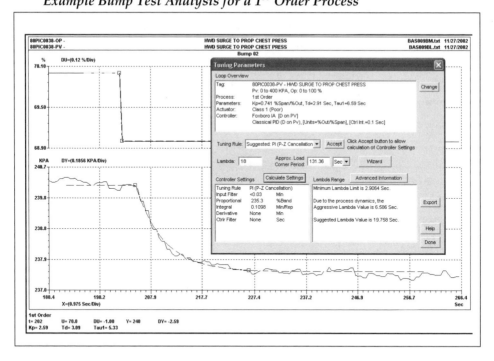

FIGURE 4-20.
Example Bump Test Analysis for an Integrating Process

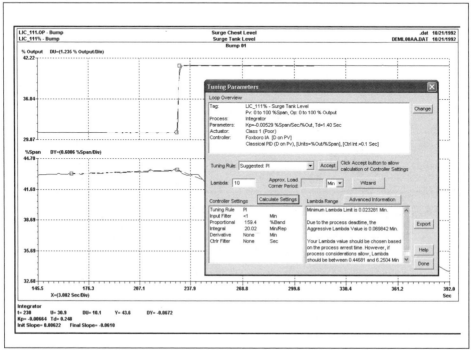

Figures 4-19 and 4-20 show how each individual bump is evaluated by fitting a curve over the actual bump and response data. The software package then helps calculate the process parameters (such as deadtime, process gain and time constant) and recommends tuning settings based on the vendor controller model. (Note: You can still perform this same analysis even if the plant does not have a software package.) Figures 4-11 and 4-12 in the previous section on bump test analysis show the calculations that should be considered and how to get the information from a trend plot.

For this example, no additional tuning on the drum level or feedwater loops is needed and no further recommendations for improvement are identified.

Reporting the Test Results

After analyzing the time series and bump test data, the graphs and narrative write ups are included with notes from the Loop Inspection data sheet and instrument HART® data in a report that summarizes the benchmarking and any recommendations for improvements.

The technician requests a meeting with Operations, Process Control, and Maintenance (and other interested personnel) and present the testing results that show the loops operating at acceptable performance expectations to meet power house objectives of stable, low-cost steam supply to the plant.

REFERENCES

1. Blevins, Terrence L., McMillan, Gregory K., Wojsznis, Willy K., Brown, Michael W. *Advanced Control Unleashed*, ISA – The Instrumentation, Systems, and Automation Society, 2003.

2. Bialkowski, W. L. *Process Control for Engineers*, Emerson Process Management, EnTech Performance Group, 2001.

3. Gerry, John. *Prioritizing and Optimizing Problem Loops using a Loop Monitoring System*, ISA – The Instrumentation, Systems, and Automation Society, 2002.

4. Fitzgerald, Bill. *Control Valves for the Chemical Process Industries*, McGraw-Hill, 1995.

QUIZ

1. What are the reasons for the pre-audit interviews?

2. What are the two insights into the process gained from the open loop bump test?

3. Why would a loop testing and analysis tool be helpful?

4. What is the most likely source of a loop performance problem and why?

5

SUSTAINING THE PERFORMANCE

The control loops are now operating at top performance, thanks to your work in the last chapters. The operators keep the loops in automatic. Operations Supervision likes the savings they're getting from reduced variability and on-spec product and, as instrument technician, you're called upon less to "fix the loop." However, as mentioned in the last chapter, since the control loop performance will degrade over time due to process changes, valve performance issues, loop tuning changes, etc., it is beneficial to monitor the loop performance.

It has been found additional benefits from implementing a loop performance program can be realized in maintenance and operations cost savings. So, we'll broaden the topic of the loop performance program to include advanced maintenance strategies and increase operator efficiencies as part of the program. Let's look at how this might apply.

5.1 MAINTENANCE STRATEGIES

Before discussing the actual aspects of a loop performance program, let's review the different maintenance strategies in a typical plant and how this might affect your program. Plant maintenance philosophies have been described by many resources (see References 1 and 2) and could be summarized in the following categories:

- *Reactive Maintenance* – Described as "fix it when it breaks," this is the most basic maintenance strategy. Its major drawback is obvious: the cost to repair (or replace) equipment that's run to failure is typically much higher than if the problem were detected and fixed earlier, not to mention the cost of lost production during extended downtime.

- *Preventive Maintenance* – A preventive strategy assumes equipment is relatively reliable until, after some period of time, it enters a "wear-out" zone where failures increase. To postpone this wear-out, equipment is serviced on a calendar- or run-time basis – whether it needs it or not. On average, this "fix it just in case" approach is about 30% less expensive than reactive maintenance. But, this approach sometimes applies maintenance too soon, can create problems in the equipment and may be too late after the failure has occurred. In fact, about 30% of preventive maintenance effort is wasted, and another 30% is actually harmful (see Reference 3).

- *Predictive Maintenance* – The third strategy overcomes these drawbacks by constantly monitoring actual equipment condition and using the information to predict when a problem is likely to occur. With that insight, you can schedule maintenance for the equipment that needs it, and *only* what needs it, before the problem affects process or equipment performance.

- *Proactive Maintenance* – The next strategy is *proactive* maintenance, which analyzes why performance is degrading and then corrects the source of the problems. The goal is not just to avoid a hard failure, but to restore or even improve equipment performance. For example, a valve failure might be caused by excess packing wear, which in turn was caused by poor loop tuning that caused the valve to cycle continuously. Retuning the loop will prevent further failures while also improving process performance.

A "best-practices" plant uses predictive maintenance for most equipment where condition monitoring is practical. This limits reactive and preventive strategies to equipment that's not process-critical and would cause little or no collateral damage if run to failure.

It could be said a loop performance program is one step toward a predictive maintenance strategy in that we are monitoring the actual condition of the critical loop components and using appropriate evaluation criteria that can begin to predict when a problem may occur. Other non-critical loops can be assigned to the reactive or preventive maintenance categories.

5.2 OPERATOR EFFICIENCY

Operators typically have extensive real-world knowledge of the plant and the process. But instead of using this know-how to improve operations, they spend much of their time and talent reacting to unexpected situations – a productivity drain that limits the number of loops they can manage effectively.

This productivity problem often begins with the loop – instruments, valves, tuning and process equipment that don't perform as they should, requiring intense operator intervention to maintain control.

Again, a loop performance program is part of a plan to address the time spent by operators reacting to troublesome loops.

The following discusses some important aspects of a loop performance program that will help you sustain loop performance and properly maintain equipment before product specifications and/or unplanned shutdowns are affected.

5.3 WHERE TO START?

As mentioned in Chapter 1, 60% of loops are not performing the intended function due to problems with valve and transmitter maintenance and tuning. The highest return on your "control system investment" comes from field device performance and tuning improvements. Thus, it would seem to follow a loop performance program should include monitoring each piece of the loop from transmitter to PID algorithm to final control element.

5.4 SELECTING THE SCOPE OF YOUR LOOP PERFORMANCE PROGRAM

So how do you decide which loops to monitor in your program? We have seen a number of different methods that usually result from different plant personnel and philosophies of maintenance/operations. Here are a couple suggested approaches:

- *Talk with Operations* – Usually the operators know which loops are causing them the most problem, and operations management will have a good idea of their "money-maker" loops. This list could be a good place to start since, if you please the operators, you're a good ways toward achieving a successful program. Also, maintenance and engineering could weigh in on their choices.

- *Unit Operation Key Loops* – if you have an area of the plant selected for your program, take each unit operation and list the loops that "feed" the unit and the "energy" loops (e.g., steam, heated fluids, etc.) within the process as a starting point. If the feed loop(s) are swinging due to poor loop control, then the rest of the process also can confront problems. Energy loops can cause off-spec product. If improved, they can provide economic savings.

FIGURE 5-1.
Example Selection of Loops Showing Economic Performance (Courtesy of ExperTune Inc., www.planttriage.com. © 2003-2004 ExperTune Inc.)

Biggest payback loops

Loop name	Unit operation	Description	Average economic assessment	Snooze loop
LIC100	Condensate Unit	Knockout Drum Level	35.6%	Snooze
FIC100	Condensate Unit	Condensate Return	25.5%	Snooze
FIC101	Boiler Feed Preheater	Preheat Steam Condensate Flow	17.7%	Snooze
TIC101	Boiler Feed Preheater	Preheat Temperature	14.6%	Snooze
PIC100	Condensate Unit	LP Steam Makeup	2.0%	Snooze

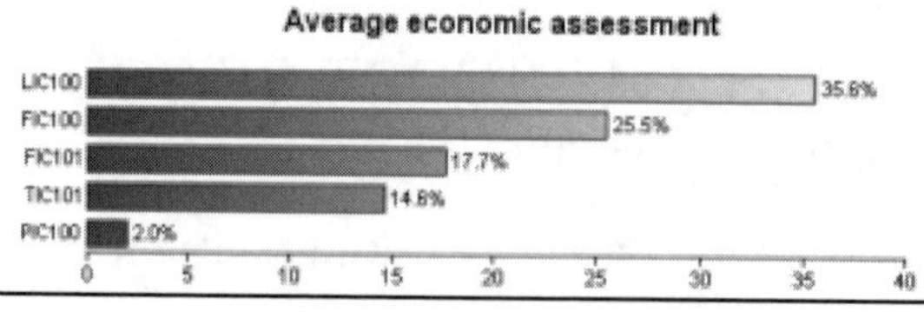

Average economic assessment:
- LIC100: 35.6%
- FIC100: 25.5%
- FIC101: 17.7%
- TIC101: 14.6%
- PIC100: 2.0%

- *Loop Rating System* – by collecting data from all loops in a plant or unit operation, and then ranking loops according to various "performance rating" calculations or "key performance indicators" (KPI), certain problem loops will begin to surface that should then be looked at for improvement. Care should be taken when ranking these loops so you are truly focusing on the key parameters of your plant's economic performance. For example, some tank levels are designed as a buffer against upsets and for process mixing. Thus, deviations from set point are not as critical as, for example, an expensive chemical flow loop.

5.5 LOOP PERFORMANCE MONITORING/ANALYZING

Once you have selected the loops for your loop performance program, the next step is setting up for monitoring and analyzing each element of the loop – transmitter, PID algorithm, valve etc. Implementing a preventive/predictive maintenance routine for the critical loop components will increase the loop performance and contribute towards your goal of reduced maintenance and increased operator efficiencies. Also, there are significant advantages to smart instruments and associated diagnostics, analysis and reporting packages that we'll discuss in the following paragraphs.

Transmitters

The old saying that "you can't control what you can't measure" applies here because the control is only as good as measurement performance. Your measurement should be accurate and repeatable and, therefore, requires periodic calibration to assure performance specs are maintained. While you're making the trip to the transmitter, look for things like high ambient temperature, verify heat tracing – on in winter, off in summer, plugged impulse lines, isolate the transmitter and vent to check zero and note any drifting that may require replacement, and empty pipe and grounding strap checks for mag meters. Over-pressuring the transmitter can also affect the performance over time. Work with your measurement vendor to establish the list of performance checks that should be accomplished, based on the measurement type. Basically, you'll want to check calibration and calibration history, verify operating data sent to the control system is accurate, and check any problems that may be affecting the transmitter.

Traditional Transmitters

For traditional transmitters (non-smart), you will want to set up a periodic preventive maintenance event that includes installation checks and a calibration to assure the device is performing as desired. Keeping good records of the periodic checks is important to analyze problems that may continue over time, and also to set the time interval between checks. For example, if the transmitter calibration is within spec after several six-month checks, then you may want to lengthen the time between checks to one year. Verifying the operating data may be limited to gauges or local reading of the process and comparing with the 4–20 mA output. As you can see, determining the overall performance health of the transmitter is somewhat limited with just the 4–20 mA signal, compared to the newer smart versions available that can communicate additional information.

Smart Transmitters

The smart transmitter helps this process by providing on-line diagnostics which, along with auto-documenting features in associated software packages, will streamline your analysis, calibration and record keeping. By connecting to the transmitter (even while online), the technician is able to gather significant diagnostic data to determine the performance health of the transmitter, while also providing configuration and operating data.

Figure 5-2 shows the results of a typical calibration check that is combined with similar checks, over time, to roll up into Figure 5-3. The calibration results, shown graphed over time in Figure 5-3, will help to determine the frequency for future checks. This could be done by hand. However, you've seen some examples from an asset management software package to pass along ideas for your plant.

FIGURE 5-2.
Example of a Typical Calibration Summary

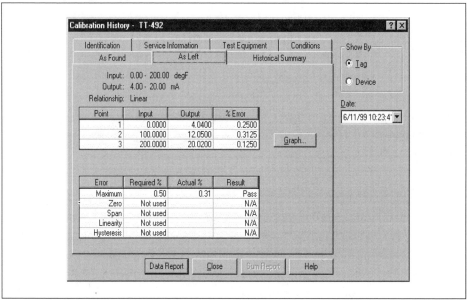

FIGURE 5-3.
Example of a Calibration History

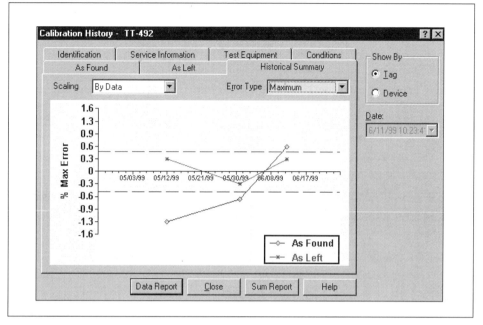

In addition to the calibration history, Figure 5-4 shows an example of temperature transmitter operating data with an alert that a diagnosis indicates the primary sensor has failed and the unit is running on the backup. Again, we're using a software package as an example, but similar data could be collected by hand-held communicators and smart transmitters.

As shown previously, a smart transmitter with an asset management software package brings considerable helpful data and convenience to the program of monitoring the loop performance. By connecting to the transmitter, the above information can be captured on a "snapshot" basis. The next step is on-line monitoring of the transmitter status that alerts you to a potential problem. An example of an alert summary from an asset management software package is shown in Figure 5-5.

In addition, there are some control systems with the ability to route these alerts to the operator as well as the maintenance shop. As we discussed earlier, operator efficiency is an important consideration. By knowing potential problems in loop devices, the operator can decide on proper action to take and request technician help to resolve the problem before a shutdown or off-spec product occurs. The operator notification example below in Figure 5-6 does not contain the detail the maintenance technician receives at the asset management software level, but it does allow the operator to make an informed decision.

FIGURE 5-4.
Example of Operating Data Display Showing Unit Is Running on Backup

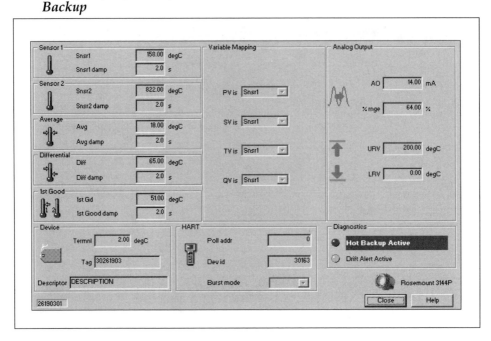

FIGURE 5-5.
Example of On-Line System Alert Summary

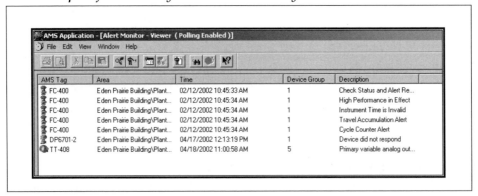

Loop Checking 97

FIGURE 5-6.
Example of Operator Interface Display with Hardware Alert

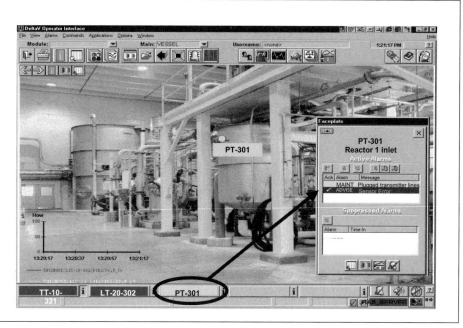

The above on-line features for your loop performance monitoring system not only require the smart field devices but also communication protocols, asset management software packages specifically designed for instrument predictive maintenance, and interface options to the control system. Later in this chapter, we will discuss the different control system architecture options available that allow this level of performance monitoring.

PID Algorithm

Once the transmitter variable is verified as performing satisfactorily, and is being monitored, the next element in the control loop is the "PID algorithm" and control strategy. Although tuning methods and control strategy design are outside the scope of this guide, you will still want to have performance evaluation methods to track for the loop performance program monitoring and be able to present this information to the operator or loop performance team members.

Some control systems will contain loop performance monitoring capabilities. Or, you can use external PCs to gather data from the control system through communication standards such as OLE for Process Control (OPC) and calculate the important loop metrics. Figure 5-7 shows an example of a control system with built-in performance data calculations (such as standard deviation). For systems without the benefit of built-in or add-on performance monitoring software, you may have to set up a manual data-entry system using the trends and operator logs, plus periodic visits to the control room for snapshot data.

FIGURE 5-7.
Example of Control System Loop Performance Data

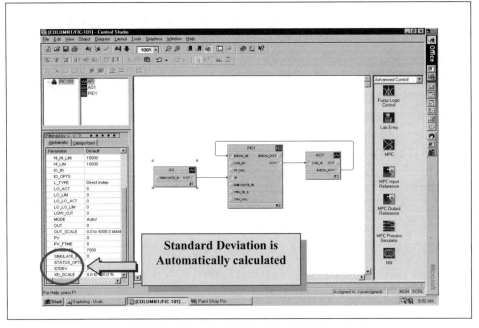

As mentioned in the last chapter, there are a number of statistical calculations for loop assessment that your software package or spreadsheet can make – such as Harris Index, integrated actual error, variability, etc. – that you'll want to choose to monitor your loops on an on-going basis. Below are some additional examples of general performance evaluation parameters that might be included in your PID performance monitoring package.

Incorrect Mode

This is a good indicator of how the loop is performing from the operator's point of view. This parameter calculates the percent of time during the day the loop is not in automatic (or cascade, depending on loop mode types).

Limited Control

If the controller output stays near 0 or 100%, then there is a problem with sizing or process condition changes that needs to be investigated. This parameter calculates the amount of time during the day the loop output has been limited.

Uncertain Input

If the control system has capabilities to recognize off-scale or uncertain inputs, or can communicate with the smart transmitters for diagnostics, then this parameter could identify other process problems or failing transmitters. This calculation is the amount of time during the day the input has been marked uncertain.

Large Variability

Although large is relative to the loop, and should be obtained from historical data, this parameter can tell you if the controller is unable to maintain set point in the desired range. Various statistical techniques can be used for this calculation. The parameter shows the amount of time the loop variability has exceeded your target for the last day.

Economic Assessment

By incorporating cost of raw materials, downtime or off-spec product, you can calculate relative economic benefits of the loop variability. The higher the economic assessment indicator the greater the benefit to your plant you will achieve by improving the loop variability.

The above loop evaluation parameters could be compared to the "dashboard" indicators in your car that tell you that something is wrong but you need more in-depth troubleshooting to find out the exact cause of the problem. Figures 5-8 and 5-9 are examples of displays that have been used to present the loop performance data.

Final Control Device

We know that the final control devices, especially control valves, are the main source of performance degradation. By keeping the valve calibrated and checked for performance, some field device problems can be avoided before the above-mentioned PID performance parameters are affected. However, a large percentage of valves tend to fall in the reactive maintenance mode. This is because most valves will continue to work even as the overall performance is decreasing; plus, it can be difficult to determine the valve's performance status. In addition, there are no

FIGURE 5-8.
Example of Loop Economic Performance Summary (Courtesy of ExperTune Inc., www.planttriage.com. © 2003-2004 ExperTune Inc.)

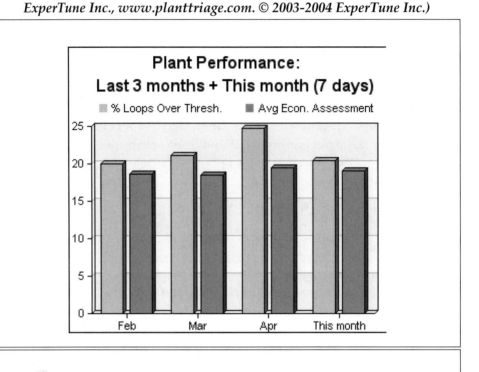

Location	1 Month	2 Months Ago	3 Months Ago	4 Months Ago	5 Months Ago	6 Months Ago	7 Months Ago	8 Months Ago	9 Months Ago
Entire Plant	24.9%	35.6%	32.8%	38.9%	46.2%	52.8%	68.9%	56.2%	65.6%
Crude Unit	22%	20%	12%	25%	10%	29.8%	28.4%	32.2%	29.5%
Sat Gas	40%	50%	52.8%	63.9%	56.2%	35.8%	74%	58.2%	70.6%
Cat Crack	38%	44%	42.8%	30.9%	36.2%	42.8%	55.9%	66.2%	68.6%

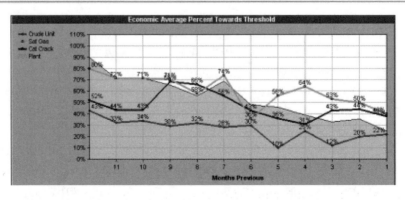

FIGURE 5-9.
Example Display Showing Loop Performance Data

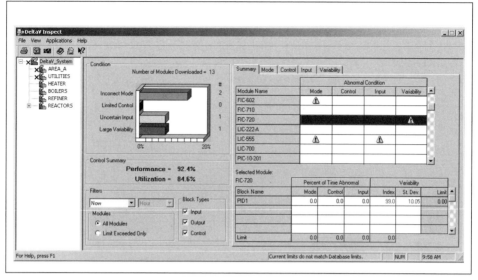

standards that are widely accepted to judge valve performance. Thus, most valves are run to failure or are identified for periodic rebuilds and quick calibration checks. The following are some examples of valve preventive/predictive maintenance that can be applied to your performance monitoring program.

Traditional

For traditional valves (non-smart), you will want to set up a periodic preventive and predictive maintenance program for the critical valves. The preventive program would include mechanical checks and calibrations to assure the device is performing as desired. Working with your valve vendor, develop a good list of the checks and adjustments that should be made. This will include such items as checking diaphragm casing bolts, benchset, fitting and tubing leaks, air supply, zero and span adjustments.

However, because it is difficult to determine the valve's performance from a preventive maintenance check, some plants will take the valve out of service and inspect or re-build on a periodic basis. This is costly, time consuming, may not be necessary, and may even introduce problems during the teardown and rebuild. To help with this effort, there are predictive maintenance tools offered by valve vendors that use sensors attached to the valve which are connected to a computer for high-speed and high-resolution data gathering, and resultant analysis of a valve's performance, while still in the pipeline. Although the valve has to be out of service (usually performed during a outage), you get a baseline of the valve performance that can be used over time to determine when actual removal and repair is needed. Figure 5-10 shows a typical setup of a valve predictive maintenance tool.

FIGURE 5-10.
Example of a Valve Diagnostic Tool

These extended valve performance diagnostics include tests and documentation for valve and I/P signatures, dynamic error band, average friction, seat load, benchset and other performance indicators (see Figure 5-11). Since there is probably not much historic data on valve failure rates and types, you'll want to keep good records on these periodic checks to understand the possible predictive techniques that could be used and how often the valve performance checks should be performed.

FIGURE 5-11.
Example of Off-Line Valve Diagnostic Test

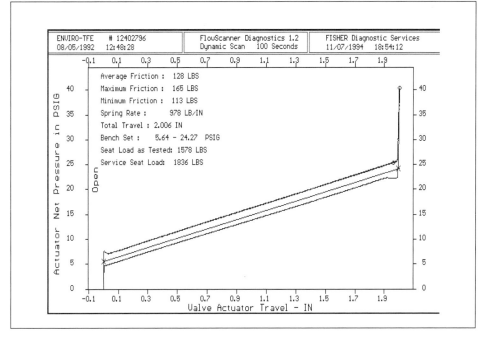

Smart Valves

The introduction of "smart valves" (typically, integrating a microprocessor into the positioner that can also communicate with a standard protocol) has enabled preventive maintenance checks during on-line operation of the valve, along with diagnostics and auto-documenting features in associated software packages that can aid in predictive maintenance. By communicating with the valve while online, the technician is able to gather diagnostic data, similar to the off-line packages described above, along with configuration and operating data.

By comparing this data over several time periods, along with the built-in diagnostics help from the vendor software, the technician can predict when the valve will need maintenance – preventing the problem from affecting overall loop performance and product quality, or causing a unscheduled shutdown.

FIGURE 5-12.
Example of an On-Line Valve Diagnostic Check

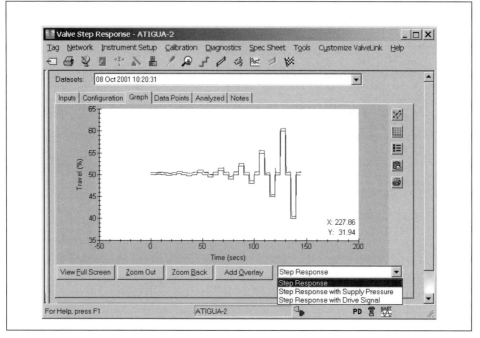

5.6 PERFORMANCE REPORTING

With the above on-line loop and field device performance data, you're now ready to keep operations and plant management informed of the control and financial performance of the control system. For day-to-day operations, take advantage of any performance display and alert capabilities in the control system. Pull together each loop's key performance data along with field device status in tabular form in a report. Using simple green, yellow or red color-coding can call attention to good, questionable or immediate attention loops. You could cut and paste from your various analysis packages into a word processor document, as shown in Figure 5-13, or use loop performance software packages that typically have good reporting capabilities, with some including E-mail capability for the reports at the touch of a button (see Figure 5-8).

Operations Supervision and Process Engineering should be very interested in your results and may ask for an update at their meetings. Weekly reports may be good at first, with monthly reports becoming more appropriate as changes in performance start to plateau. In each report, remind the report recipients of the savings you're delivering. Trends over time of performance indicators make a good visual to add to the report.

Be sure to make extra effort to inform all interested parties when your performance monitoring system catches a loop problem – before the product spec is affected or downtime occurs.

FIGURE 5-13.
Example Word Processor Report

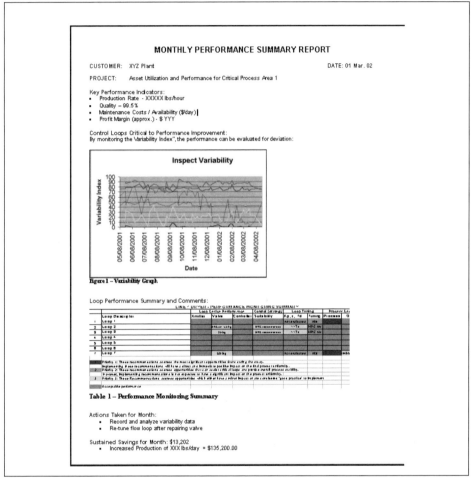

5.7 LOOP PERFORMANCE PROGRAM ARCHITECTURE

We've looked at the main areas of the loop performance program – including the transmitter, PID algorithm and control valve – and have discussed the parameters that should be monitored. Gathering the appropriate data, analyzing the data for problem detection, calculating the performance criteria, generating summary reports and alerting the proper people can be a large task. Automating some or all of these functions can help you implement a smooth running program. There are various options available for your control system to build toward an architecture that will allow the desired performance monitoring that you need.

Traditional

Let's start with a "traditional" system of non-smart field devices (4–20-mA signals) wired to a control system I/O structure, process controllers that perform the PID control algorithm and communicate the information to human-machine interfaces (HMIs) for operator control (see Figure 5-14). In this case, most of the performance program monitoring/reporting and analysis will be based on manually collected data from the field devices and control system trends. Spreadsheets or other tools are then used to calculate and report performance. The field device performance monitoring will consist of the calibration visits, checklists and valve re-builds or diagnostic tests during an outage.

FIGURE 5-14.
Example of a Traditional Control System

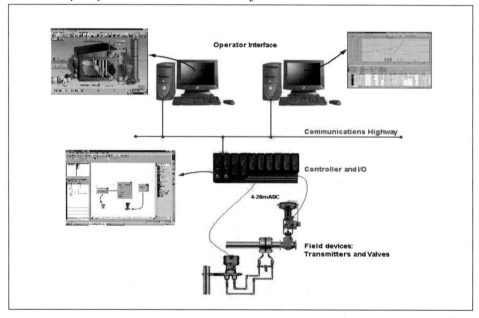

Smart Field Devices & Loop Performance Software

The next step might be to add a loop performance software package and smart field devices to your control system, as shown in Figure 5-15. The loop performance software package would collect data from the control system (using OPC or other data transfer methods) and automate the calculations that will be used to evaluate each loop and provide the reporting capability. A portable communicator and/or laptop PC) is used to "talk" with each smart instrument. This gives you an idea of operating data confirmation and some diagnostics (although on a snapshot basis, when you make the communication connection).

FIGURE 5-15.
Adding Loop Performance Software and Smart Field Devices

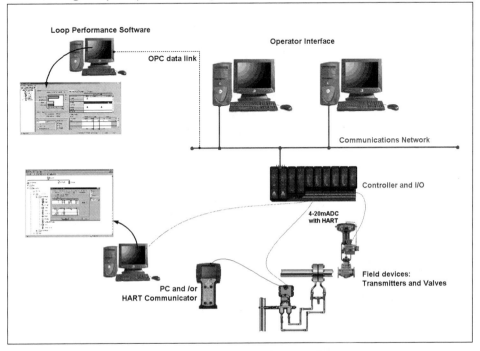

On-line Device Monitoring

To get the full benefit of the predictive capabilities of the smart instruments and software packages, you'll want to go on-line with the field devices – constantly scanning for alerts and giving you easy access to the field devices. Figure 5-16 is an example of how this might be done by adding a smart device interface to the control system and "jumpering" the smart device data to the interface. The smart field device interface then communicates to a field device asset management software package that is continually looking at the non-control diagnostic and alert information from the field devices, while the 4–20 mA control signal passes directly onto the control system. Also, the computer for running the field device management software can reside in the instrument shop or appropriate location for easy access. This minimizes time for troubleshooting and trips to the field (and hazardous environments).

With this architecture, you now have on-line monitoring of the loop performance from transmitter to PID algorithm to valve.

FIGURE 5-16.
Adding On-Line Field Device Monitoring and Alerts

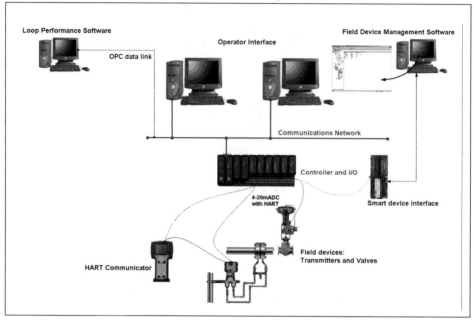

Integrated Performance and Control System

The system in Figure 5-16 might seem a bit difficult to manage with the different devices from multiple vendors, multiple databases and connection protocols. There are some vendors that incorporate loop performance data and field device monitoring/alert into the control system such that the architecture looks like Figure 5-17. Then, the control and device performance data (from a variety of smart field devices such as HART® and FOUNDATION™ Fieldbus) is available to the operator or maintenance personnel at the operator interface.

FIGURE 5-17.
Integrating Loop Performance and Field Device Management with Control System

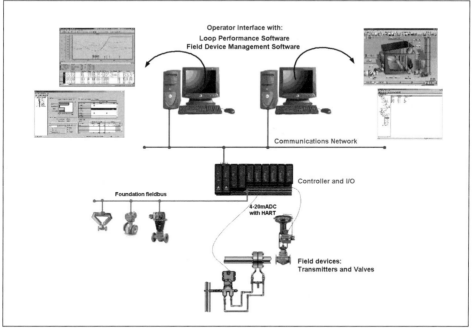

5.8 LOOP PERFORMANCE PROGRAM SUMMARY

You now have a loop performance program in place and running that is monitoring all critical loops and alerting the proper maintenance and operations personnel when an element of the loop begins to show a problem – before product quality or unplanned shutdowns might occur. The predictive maintenance aspects of your program begin to save maintenance costs and the operators are more efficient in their job. You're able to report the savings that are being made. Continuing improvement programs at your plant may start to look at the loops to see what further improvements can be made or advanced control packages can now operate without constraints from under-performing base level control. Some have even looked to loop device performance specifications as the main purchasing criteria.

This same loop performance program concept of
- evaluating performance,
- making improvements to maximize performance, then
- monitoring over time to sustain the performance through predictive maintenance techniques

is also being applied to other areas of the plant, such as rotating equipment, process equipment such as boilers, turbines, and unit operations.

The loop performance program is a continuing process. As loop problems are found with monitoring, you'll return to Chapter 4 to start the loop check process to find the problem, make the fix, and then return the loop to monitoring. As additions to the control system or new projects are added to the unit operation, this manual can be used to check the loop from factory acceptance test through start-up then into performance benchmarking and monitoring.

REFERENCES

1. *Reducing Operations and Maintenance Costs*, Emerson Process Management, 2003.

2. Fitzgerald, Bill. *Control Valves for the Chemical Process Industries*, McGraw-Hill, 1995.

3. Berlanger, Dennis and Saxon Smith. *The Business Case for Reliability*, as published at www.reliabilityweb.com/rcm1, MRG Inc.

4. Paulonis, Michael and John Cox. *A practical approach for large-scale controller performance assessment, diagnosis and improvement*. Journal of Process Control, 2003.

5. Gerry, John. *Prioritizing and Optimizing Problem Loops using a Loop Monitoring System*, presented at ISA Expo 2002, ISA – The Instrumentation, Systems, and Automation Society, October 2002.

6. "DeltaV Inspect and how it reduces Process Variability", Emerson Process Management, 2002.

7. Johnson, Mike and Joe Bilodeau. *Baby Steps to a World Class Valve Maintenance Program*, Weyerhaeuser-Canada, 2004.

QUIZ

1. What two advanced maintenance strategies are employed by loop performance programs?

2. How does the operator benefit from a loop performance program?

3. Why are smart field devices (transmitters and valves) desirable for a loop performance program?

4. My plant does not have smart field devices or a DCS. Do I have to upgrade (and spend many dollars) so my controllers are able to start a performance monitoring program?

ACRONYMS

ACC	American Chemistry Council
AIC	Availability, Integrity, and Confidentiality
AIChE	American Institute of Chemical Engineers
AWWA	American Water Works Association
BCIT	British Columbia Institute of Technology
BPCS	Basic Process Control System
CCPS	Center for Chemical Process Safety
CIDX	Chemical Industry Data Exchange
CIO	Chief Information Officer
CISA	Certified Information Systems Auditor
CISSP	Certified Information System Security Professional
COTS	Commercial Off The Shelf
DCS	Distributed Control Systems
DHS	Department of Homeland Security
DoE	Department of Energy
FERC	Federal Energy Regulation Commission
GAO	General Accounting Office
HMI	Human Machine Interface
M&CS	Manufacturing and Control Systems
NERC	National Electrical Reliability Council
NIST	National Institute of Standards and Technology
NISCC	National Infrastructure Security Co-ordination Center
NRC	Nuclear Regulatory Commission

OCIPEP	Office of Critical Infrastructure Protection and Emergency Preparedness
OPC	Object Linking and Embedding for Process Control
PCSRF	Process Control Security Requirements Forum
PLC	Programmable Logic Controllers
SCADA	Supervisory Control and Data Acquisition
SIS	Safety Instrumented Systems
SPDS	Safety Parameter Display System
TCP/IP	Transmission Control Protocol/Internet Protocol

Appendix A
TUNING

The word *loop* is often used in discussion about process control. A loop, the basic unit of process control, typically consists of one process measurement, one set point, and one output.

A process usually has more than one loop, and many processes used in industry have thousands of loops. Some of these loops are independent of other loops; some loops are interconnected with other loops.

A variable that is measured by the process instrumentation is usually known as the measured variable or process variable. Examples of measured variables are liquid and gas flows, levels, temperatures, and pressures.

Some variables are not directly measured from the process but are calculated by the control system (usually using measured variables). Examples of calculated variables are differential temperature (calculated by subtracting two measured temperatures), and a flow ratio (calculated by dividing two measured flows).

LOOP CLASSIFICATION BY CONTROL FUNCTION

Each loop has a measured (or calculated) variable. These variables are usually measurements made from the process using sensors or transmitters. Variables can be classified based on how they are used in the control system.

Controlled Variable

A controlled variable is held by the control system at either a target or set point. The primary function of a control system is to control these variables.

Manipulated Variable

For every controlled variable, at least one must be manipulated, that is, directly adjusted by the controller or the operator. The manipulated variable is usually a flow that passes through a control valve or some other final control element. It may or may not be measured.

Indicated Variable

Many variables are measured but are not controlled or manipulated. These are indicated only for operator information or for record keeping, but no effort is made either by the control system or by the operator to keep that variable at a particular value.

In Figure A-1 the level of the tank is a controlled variable; it is desired to hold this variable near a set point to prevent the tank from emptying or overflowing. To *control* the level, the flow out of the tank is *manipulated*. Other than for the control of the level, there is reason for requiring a particular discharge flow rate. It may be measured for indication or recording, but this measurement is not needed for control.

FIGURE A-1
Level Control

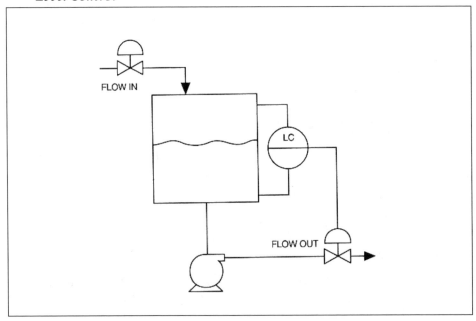

Flow loops usually contain both controlled and manipulated variables. In a few cases the flow that is manipulated is not the same as the controlled flow. For example, the total flow in Figure A-2 is the controlled variable, while flow A is the manipulated variable.

The terms manipulated, indicated, and controlled variables include not only variables that are directly measured but also those that are inferred: variables that are calculated based on other inputs. In Figure A-3 a ratio between two flows is manipulated to control the composition of the blended flow.

In Figure A-4 a differential temperature within a process is to be maintained. The controlled variable is the differential temperature; that is, it is inferred from two individual temperatures.

FIGURE A-2
Controlled and Manipulated Variables

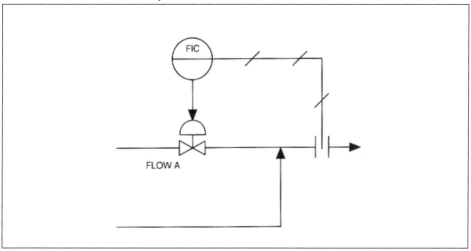

FIGURE A-3
Manipulated Inferred Variables

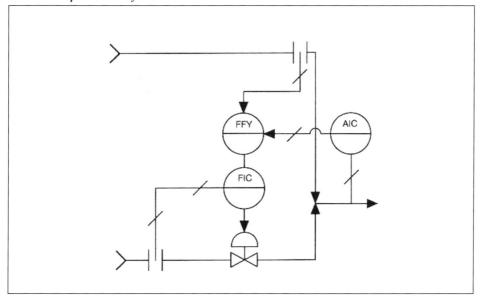

FIGURE A-4
Inferred Controlled Variables

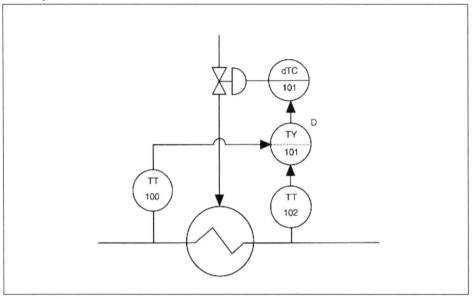

Set Point

Each control loop has a set point—the desired value for the controlled variable. For example, the set point for the temperature of a room, as set on the thermostat, may be 70°F. The set point may remain constant or may be often adjusted by the operator, by other control loops, or by a computer program.

CONTROL ALGORITHMS

The controller compares the measured variable to the set point and determines an output to the manipulated variable. To make this determination, an "algorithm" is used, which is a step-by-step procedure for solving a problem. Explanations of some of the simple basic algorithms follow; then the PID algorithm used in industrial control is examined.

Basic Algorithms

The most basic form of control algorithm is a simple on/off control. A switch detects whether the process variable is above or below the set point. The state (on or off) of the manipulated variable is changed when the controlled variable crosses the set point.

This type of control is used successfully in many situations. Most home appliances use on/off control to switch the power to the heating element on and off to maintain the desired temperature.

Because on/off control often tends to cause the manipulated variable to be turned on and off too often, a variation of the on/off control adds a gap or dead band in the comparison of the controlled variable with its set point. This causes the manipulated variable to be changed less often but also increases the amount of variation in the controlled variable.

The PID Algorithm

The most common algorithm used in industrial process control (almost the only algorithm used) is the time-proven proportional-plus-integral-plus-derivative (PID) algorithm.

The PID control algorithm does not "know" the correct output to bring the process to the set point. It merely moves the output in the direction that should move the process toward the set point. The algorithm must have feedback (process measurement) to perform.

 The PID algorithm must be tuned for the particular process loop. Without such tuning, it will not function. To properly tune a PID loop, each of the terms of the PID equation must be understood. The tuning is based on the dynamics of the process response.

Note: The controller action is always the opposite of the process action.

Most control systems in industrial process control allow the operator to place a loop into automatic or manual control. Manual mode allows the operator to manipulate the output to bring the measured variable to the desired value. This mode is often used during the start-up of the process (see Figure A-5). In automatic mode the control loop manipulates the output to hold the process measurements at their set points. This should be the most common mode for normal operation. See Figure A-6.

The most important configuration parameter of the PID algorithm is the action. This determines the relationship between the direction of a change in the input versus the resulting change in the output. Direct action means that an increase in input results in an increase in the output. Reverse action means that an increase in the input results in a decrease in the output.

FIGURE A-5
Manual Mode

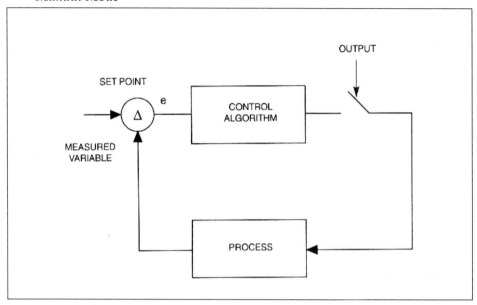

FIGURE A-6
Automatic Mode

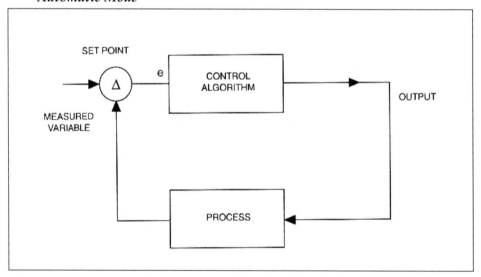

The PID controller, as the name implies, is made of three basic responses: proportional (or gain), integral (or reset), and derivative. This section presents the individual responses that make up the PID controller.

The error is the difference between the process and the set point. If the controller is direct-acting, the set point is subtracted from the measurement; if it is reverse-acting, the measurement is subtracted from the set point. Error is always in percent.

Error = Measurement – Set point   (Direct action)

Error = Set point – Measurement   (Reverse action)

PROPORTIONAL

The most basic response is proportional, or gain, response. In its pure form, the output of the controller is the error times the gain added to a constant known as manual reset.

$$\text{Output} = G \times e + K$$

where *G* is gain, *e* is error, and *K* is manual reset.

If the manual reset remains constant, there is a fixed relationship between the set point, the measurement, and the output.

The output of a proportional-only controller always follows the error, with no dynamic (time-based) difference.

When there is a disturbance (or load change), proportional-only control reduces but does not eliminate the error. The remaining error is known as offset.

The tank in Figure A-7 has liquid flowing in and flowing out under control of the level controller. The flow into the tank is independent and can be considered a load by the level control. The flow out is driven by a pump and is proportional to the output of the controller.

Assume first that the level is at its set point of 50%, the output is 50%, and both the flow in and the flow out are 500 gpm. Then assume the flow in increases to 600 gpm. The level will rise because more liquid is coming in than going out. As the level increases, the valve will open and more flow will leave. If the gain is 2 for every one percent increase in level, the valve will open 2% and the flow out will increase by 20 gpm. Therefore, by the time the level reaches 55% (5% error) the output will be at 60% and the flow out will be 600 gpm, the same as the flow in. The level will then be constant. This 5% error is known as the offset.

Offset can be reduced by increasing gain. Repeating the above experiment with a gain of 5, for every 1% increase in level the output will increase by 5% and the flow out by 50 gpm. The level will have to increase only to 52% to result in a flow out of 600 gpm and cause the level to be constant.

Increasing the gain will reduce the offset, but only an infinite gain will totally eliminate offset.

FIGURE A-7
  Offset

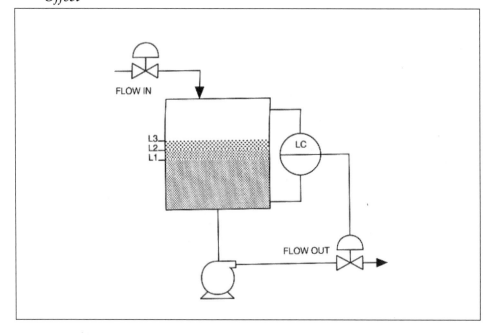

Figure A-8 illustrates the relationship between measured variable and output. With no load, a line can be drawn from the vertical (output) axis to the line L1. A line from the intersect is then drawn down to the horizontal (measured variable) axis. Likewise, a line is drawn from the measured variable axis to the control line. From the intersect a line is drawn left to the output axis.

If we switch from the line L1 to line L3 (indicating a change in the load), the line drawn from the original output to L3 and down will indicate the change in measured variable if there is no control action. Compare the offset with low proportional gain and high proportional gain.

Offset can be eliminated by changing the manual reset every time the load changes. After the load has changed, the operator can increase the manual reset to open the valve further or decrease the manual reset to close the valve. The operator continues to adjust the manual reset until the offset is eliminated.

INTEGRAL

Whenever the load changes and the operator notices an offset in the control loop, the operator "resets" the controller (to move offset) by adjusting the manual reset.

FIGURE A-8
Measured Variable/Output Relationship

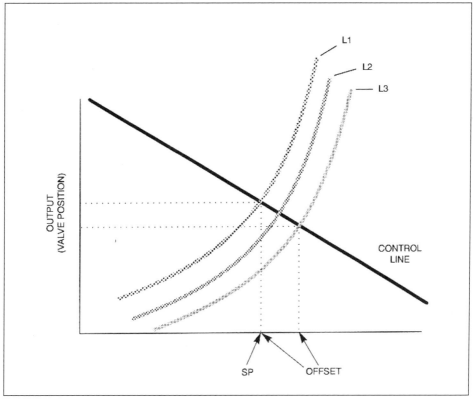

If the load does not often change or the offset is too small to create a problem, proportional-only control is sufficient. If the offset and the need to frequently adjust the loop is a problem, the manual reset may be replaced by automatic reset. Automatic reset is a function that continues to move the output whenever there is any error. Automatic reset is usually called *reset* or *integral action*. One way to add reset action to a controller is to add a circuit or function that takes the controller output, performs a lag, and adds the result to the product of the gain times the error.

Note the use of the positive feedback loop to perform integration (see Figure A-9). As long as the error is zero, the output will be held in place. However, if the error is not zero, the output will continue to change until it has reached a limit.

FIGURE A-9
Integration

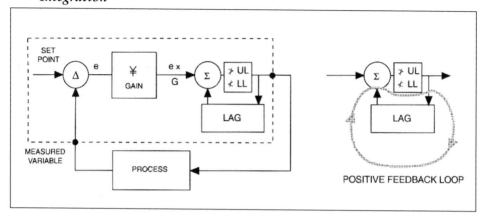

From a more mathematical point of view, the reset contribution is:

$$\text{Output} = G \times K_i \times \int e \, dt$$

where:

G = gain
K_i = reset rate, repeats per min.
e = error, %

Figure A-10 graphically represents pure reset.

FIGURE A-10
Reset

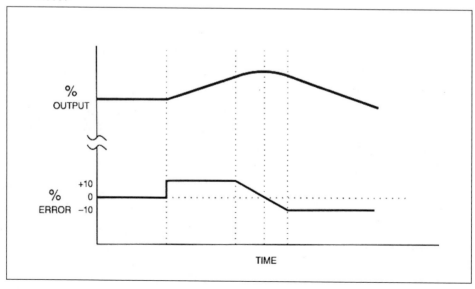

Note: Some control vendors measure reset by *repeat time* in minutes. Others measure reset by *repeats per minute* the inverse of minutes of repeat.

If one looks at the output of a controller (not connected to the process) over time following a step change in the error, two distinct effects (see Figure A-11) are obvious. Immediately the output changes by the amount of the change in the error (in percent) times the gain. This is the *gain effect*. Then the output continues to move as long as there is an error. This additional change in the output is known as the *reset effect*. The time between the change in the error and the point in time when the reset effect equals the gain effect is known as the repeat time or reset time.

FIGURE A-11
Gain and Reset Effects

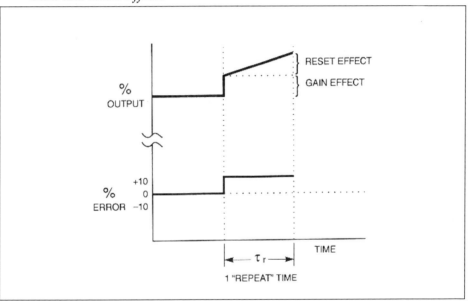

One problem with the reset function is that I may "wind up." Because of the integration of the positive feedback loop, the output will continue to increase or decrease as long as there is an error (difference between set point and measurement) or until the output reaches a defined limit.

This normally is not a problem and is a normal feature of the loop. For example, a temperature control loop may require that the steam valve be held fully open until the measurement reaches the set point. At that point, the error will cross zero and change signs, the output will start decreasing, and the steam valve will begin to throttle back.

Sometimes, however, reset windup may cause a problem. Actually, the problem is not usually the windup but the "wind down" that would be required.

Suppose the output of a controller is broken by a selector, with the output of another controller taking control of the valve. In Figure A-12 the lesser of the two controller outputs is sent to the valve. The controller that has the lower output will control the valve. The other controller is, in effect, open loop. If its error would make its output increase, the reset term of the controller will cause the output to increase until it reaches its limit.

FIGURE A-12
Use of a Selector

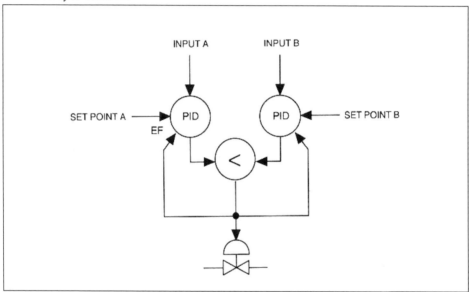

The problem is that when conditions change and the override controller no longer needs to hold the valve closed, the primary controller's output will be very far above the override signal. Before the primary controller can have any effect on the valve, it will have to "wind down" until its output equals the override signal.

The positive feedback loop that is used to provide integration can be brought out of the controller. Then it is known as "external feedback."

If there is a selector between the output of the controller and the valve (used for override control), the output of the selector is connected to the external feedback of the controller. This puts the selector in the positive feedback loop.

If the output of the controller is overridden by another signal, the overriding signal is brought into the external feedback. After the lag, the output of the controller is equal to the override signal plus the error times gain. Therefore, when the error is zero, the controller output is equal to the override signal. If the error becomes negative, the controller output is less than the override signal, so the controller regains control of the valve.

DERIVATIVE

Derivative was first used as a part of temperature transmitter to overcome lag in transmitter measurement. It is also known as rate.

The derivative contribution can be expressed mathematically as:

$$\text{Output} = G \times K_d \times \frac{de}{dt}$$

where:

G = gain
K_d = the derivative setting, minutes
$\frac{de}{dt}$ = the rate of change of the error, %/min.

The response, over time, of a controller with gain and derivative is shown in Figure A-13. While the error is changing, the output is given an additional boost ("derivative effect" in the figure). After the error has stopped changing and is constant at a new value, the derivative effect is eliminated and the only change in the output is the gain effect, just as if the controller had proportional-only control.

The output with the same gain setting but no derivative effect is shown as a broken line in the figure. If a horizontal line is drawn from any point on the broken line to any point on the solid line, it can be seen that the addition of derivative "advanced" the output by a certain time. That is, the output reaches a given value earlier if derivative is used than if proportional-only is used. The amount of time that the derivative action advances the output is known as the "derivative time," which is measured in minutes.

FIGURE A-13
Response Time and Gain

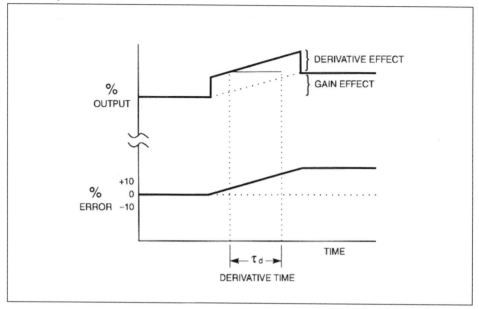

Note: All major vendors measure derivative (rate) in minutes.

COMPLETE PID RESPONSE

Combining the gain, integral, and derivative elements of the PID equation we have:

$$\text{Output} = G(e + K_i \int e\,dt + K_d \frac{de}{dt})$$

where:
  G = gain
  K_i = reset (repeats per minute)
  K_d = derivative (minutes)
  e = error, %

While any combination of the three modes (P, PD, ID, etc.) is possible, the combinations of P-only, PI, and PID are the most commonly used. At least 75% of all control loops use PI control.

LOOP TUNING

Once a loop is configured and started up, someone has to adjust the gain, reset, and derivative values in the controller. Unless these

Loop Checking

adjustments are made, the controller will not properly control the process. At one extreme, the control loop will be unresponsive, just as if the control loop were in manual. At the other extreme, the control loop will be unstable, with the manipulated variable swinging between its low and its high limits. The adjustment of these parameters is known as loop tuning.

Tuning Criteria or "How Do We Know When It's Tuned?"

One often-overlooked aspect of control loop tuning is the determination of when the loop is correctly tuned. Of course, if the loop is unresponsive to set point changes or if it is wildly swinging, it is not correctly tuned. But between these extremes it is more difficult to determine how well the loop is tuned.

Informal Methods

These methods involve observation and measurement of the measured variable after the set point has been changed or the loop has been disturbed. The choice of methods depends upon the loop's place in the process and its relationship with other loops.

OPTIMUM DECAY RATIO (QUARTER WAVE DECAY)

This common method of judging tuning is known as the "quarter wave decay" method (see Figure A-14). It has been shown that if a loop is tuned so that the oscillation decays with each wave being one quarter of the previous wave, it produces satisfactory, if not optimum, set point response and disturbance rejection.

FIGURE A-14
Quarter Wave Decay

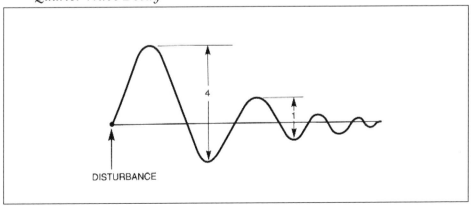

MINIMUM OVERSHOOT

For some loops the most important criterion is that the loop be able to respond to a set point change with as little overshoot as possible. For these loops the tuning will be adjusted to minimize the overshoot, even at the expense of reduced disturbance rejection (see Figure A-15).

FIGURE A-15
Minimal Overshoot

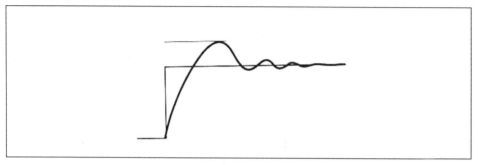

MAXIMUM DISTURBANCE REJECTION

In other cases the goal is to reduce the peak response to a disturbance even if this causes the loop to take longer to stabilize at the set point or causes an overshoot following a set point change.

Mathematical Criteria—Minimization of Index

Certain mathematical criteria numerically compare the results of a loop following a disturbance or set point change. After either a set point change or a disturbance, the error is measured and integrated using the methods listed below. After the process value has settled at the set point, the result of the integration is noted.

IAE—Integral of absolute value of error:

$$\int |e|\, dt$$

ISE—Integral of error squared:

$$\int e^2\, dt$$

ITAE—Integral of time times absolute value of error:

$$\int t|e|\, dt$$

ITSE—Integral of time times error squared:

$$\int te^2 dt$$

These mathematical methods are used primarily for academic purposes, together with process simulations, in the study of control algorithms.

What's Really Important

The real key to judging the tuning of a loop is knowledge of "what's really important" to the operation of the process. It may be important that the process be held extremely close to the set point. It may be important that the process be held in the general vicinity of the set point but with a limit of the effect of a disturbance. In other cases it may be most important that the process rapidly follow changes in the set point.

Too often loop tuning is tested by checking the set point response. It is an easy test to do: one simply makes a small change in the set point and then observes the response of the process. However, the response to set point changes is an overrated measure of loop tuning. While some loops are subject to set point changes and must respond quickly, the set points of many loops typically do not change. For these loops the set point response is irrelevant.

Experience-Based Tuning

The most common form of tuning, used by experienced instrument personnel, is performed as follows:

1. Enter an initial set of tuning constants from experience. A conservative setting would be a gain of 1 or less and a reset of less than 0.1 repeats per minute.

2. Put the loop in automatic with the process lined out.

3. Make step changes (about 5%) in set point.

4. Observe the response. Based on experience, determine the change required (if any) and adjust if further tuning is needed.

Of course, this method requires experience. Other loop tuning methods do not require an extensive knowledge base, although they are often used by experienced personnel.

Ziegler-Nichols Tuning Method, Open-Loop (Reaction Rate)

Ziegler and Nichols suggested two methods of loop tuning that are still used today. For most processes these methods will result in quarter wave decay tuning (describe above). This tuning is usually a good compromise between tuning that optimizes set point response and tuning that optimizes disturbance rejection.

The first of the Ziegler-Nichols methods is also known as the "reaction curve" or "open loop reaction rate" method. This method cannot be used with integrating processes that will not level off at some acceptable value when a small change is made in the output.

The process must be lined out, that is, stable and not changing. With the controller in manual, the output is changed by a small amount (5% is typical). The process is recorded as it reaches its new level (see Figure A-16).

FIGURE A-16
Reaction Curve

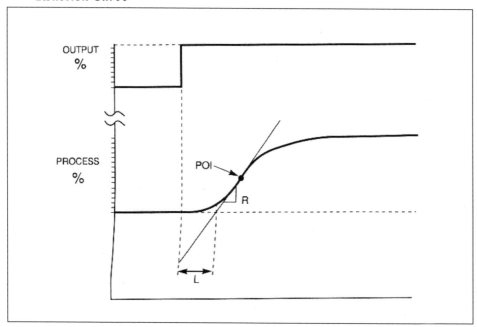

The following measurements are made from the reaction curve:

ΔX	%	Change of output
R	%/min.	Rate of change at the point of inflection (POI)
L	min.	Time until the intercept of tangent line and original process value agree (lag time)

The gain, reset, and derivative are calculated using the following:

Ziegler-Nichols Tuning Method, Closed Loop (Ultimate Period)

Another method developed by Ziegler and Nichols is the closed-loop method. In this method the loop is kept in automatic with the rest of the process in a stable condition. Unlike the open-loop method, this method can be used with integrating processes.

Steps:

1. Place the controller into automatic with low gain and no reset or derivative.

2. In steps, increase the gain and then make a small change in the set point. Watch for oscillations to start.

3. Adjust the gain to make the oscillations continue with a constant amplitude. If the oscillations tend to grow, slightly decrease the gain. If the oscillations tend to decay, increase the gain. The gain that results in constant oscillations is known as the ultimate gain (G_μ). The period, in minutes, is called the ultimate period (P_μ).

The gain, reset, and derivative are calculated using the following:

	GAIN	RESET	DERIVATIVE
P	$\frac{\Delta X}{LR}$	—	—
PI	$0.9\frac{\Delta X}{LR}$	$\frac{0.3}{L}$	—
PID	$1.2\frac{\Delta X}{LR}$	$\frac{0.5}{L}$	$0.5L$

Controllability of Processes

The "controllability" of a process depends upon the gain that can be used. As a general rule, the higher the gain, the greater rejection of disturbance and the greater the response to set point changes. Therefore, with most loops it is preferable to use the highest gain that does not result in undue oscillations.

In looking again at the response of a loop to a step change in the output, dynamic lag can be calculated. The *predominate lag* (L in Figure

A-16) is based on the largest lag in the system. The maximum gain that can be used depends upon the ratio LR. Notice the calculation for the gain in the Ziegler-Nichols open-loop method above.

From this we can draw two conclusions: (1) decreasing the dead time increases the maximum gain and the controllability, and (2) increasing the ratio of the longest lag to the second longest lag also increases the controllability.

FLOW LOOPS

Flow loops are too fast to use the standard methods of analysis and tuning.

Some flow loops that use analog controllers are tuned with high gain. This will not work with digital control. With an analog controller, the flow loop has a predominate lag of a few seconds and no subordinate lag. The scan rate of a digital controller can be considered dead time. Although this dead time is small, it is large enough when compared to the lag to require that a low gain be used.

For most flow loops, good control will be obtained with a gain from 0.5 to 0.7, a reset from 15 to 20 repeats/min., and no derivative.

The above Appendix A text was originally published in **Maintenance of Instruments & Systems, 2nd Edition** *by Larry Goettsche, Editor. Copyright (c) 2005 ISA - The Instrumentation, Systems, and Automation Society. For more information, please visit www.isa.org/books.*

Appendix B
ANSWERS

CHAPTER 1

1. Plant Manager/Profit Center manager – Follow the money. Since these people are closely tuned to the financial aspect of the business, the return on investment (ROI) is an important decision benchmark. The Dow study showed the highest ROI of nine control system "best practices" to be loop tuning and field device performance – key elements of a loop checking program.

   Project Manager/Engineer – The loop checking process, from design through the FAT and start-up, can make a significant impact on the project costs and success. Frequently, a fast, uneventful start-up is a large measure of the successful project for the project manager/engineer.

   Control System Technician – Loop checking programs can provide a structured approach to aid the technician's job of maintaining loops. Loop histories are kept which makes for better troubleshooting, results are identifiable for job performance goals, key contribution to unit operations teams.

   Process Engineer – Loop checking assures the process engineer that the control system is working correctly and doing the best job of variability reduction. He is then free to investigate other process improvement opportunities such as advanced control or process bottlenecks.

2. Field Device and Loop Tuning

3. Control Loop: measurement (transmitter) – controller – final control element (valve). The control valve is most susceptible to long-term

performance degradation due to mechanical linkages, moving parts, friction buildups, wear, lack of maintenance, and so on.

4. The loop checking process begins when instruments are received at site and during factory acceptance testing. End point – Although this guide discusses the loop check through sustaining loop performance activities, it is an on-going process.

CHAPTER 2

1. The FAT can save significant time and effort during commissioning/start-up by early problem detection and resolution before extended start-up time is required, product quality is affected or an unplanned shutdown occurs. FAT is a training opportunity for operators and maintenance personnel.

2. A system hardware check and a loop configuration check are the two basic parts of FAT.

3. By simulating the control system I/O and a plant process, Process Simulation Packages allow the control system configuration to be tested in near actual conditions. That is, I/O signals can be sent and received in software that will confirm the terminations are per the design documents; the control strategies can be confirmed since the controller "thinks" it is getting inputs and outputs from a "live" process. Simulation testing is beneficial for providing the means for more complete and accurate testing, compression of project schedules, and for on-going proof testing of new and advanced strategies. Also, on-going operator training programs can use the simulation system.

4. Smart instruments mainly come into play during a Site Acceptance Test where the instruments are installed and wired to the control system. It is time consuming for technicians to be at every instrument and input/measure signals for verification of the loop operation. By communicating to smart instruments from a central location, multiple instruments can be set up and tested through PC-based programs without disturbing wiring.

CHAPTER 3

1. To verify wiring and also that the display scaling is correct.

2. Input signals from the field device at 0%, 25%, 75%, and 100% levels, verify display at control system then set loop output at same values of 0–100% and verify at final control element.

3. Since experience has proven that significant problems are introduced by incorrect installation of field instruments and valves, the key observation by the field team member is to check for possible installation/setup problems. Note the additional capabilities that smart instrumentation provides with online diagnostics and setup.

4. Efficiency – Smart instrumentation allows for communicating additional information for checkout in addition to just the process signal. This can save trips to the field, reduce safety issue exposure from being in a process area, reduce number of people required for the checkout team, provide extensive trouble shooting diagnostics and increase accuracy of the checks.

CHAPTER 4

1. The interviews will not only give you a history of the past and current problems, and what each person is expecting, but also helps the group understand your objectives and give them some participation and buy-in on the benchmark check.

2. Process dynamics and condition of the final control element.

3. Although you could probably do some manual calculations from strip chart printouts, the loop testing and analysis tools provide faster, more accurate calculations along with additional mathematical analysis tools for simulation, tuning, cycling identification and benchmarking. The tools also provide a storage place for future review of the test data, as well as better visualization of the data for analysis and report presentation.

4. The control valve can typically be the source of the performance problem, especially if the loop has operated over a long period of

time. This is due to the mechanical wear and friction force increases that the seals, relays, packing, etc. can experience over time.

CHAPTER 5

1. Predictive and proactive maintenance techniques are key to the performance program because they predict and correct problems before failures or off-spec product occur.

2. The operator is better informed about the actual problem and can call the right resources to help or work around the problem without a lot of wasted effort. The operator's efficiency is improved.

3. Smart field devices bring added diagnostics and operating conditions, in addition to configuration/setup data, to the operator and maintenance personnel. Thus, predictive maintenance techniques can be used to help catch problems before loop variability (and thus product quality) is affected, or unplanned failures occur.

4. No, the methods described in this chapter of periodically verifying transmitter, PID algorithm and control valve performance, deciding on and calculating your plant's appropriate loop performance assessment parameters, then reporting the results to operations and maintenance supervision will work regardless of the control system type (from single loop to PLC's to DCS to hybrid to digital control systems).

INDEX

1st Order 74
2nd Order 74

AC power 29
advanced process control (APC) 1
alarm conditions 34, 36
alarms 33
alerts 95
analog 33
    input 33
    output 7, 33, 69
APC 1
asset management packages 78
audit trail 51

basic algorithms 118
benchmark data 72
bill of material 28
bump tests 61, 70–73, 76–78, 82–84, 86
business drivers 2

calibration 9, 47, 50, 60, 78, 93–95, 106
    history 95
cascade 37
    control 11, 34
configuration standards 30
control algorithms 118
control loop 1–5, 15–16, 25, 27, 33, 89, 97
    single input, single output 5
control system technician 3
control valves 17, 21, 77, 99
controlled variable 115

DC power 29
deadtime 17, 21, 70, 74
design specifications 26
discrete 38
    control devices 34–35
    control element testing 35
    input 8, 33
    output 8, 33
display testing 37
displays 34–36
distributed control system 6, 11, 25
drivers
    business 2

economic assessment 99
EnTech gain specification 19
ExperTune index 16

factory acceptance test 1, 8, 25
feedback 4, 10, 16
    control 4, 7, 11
feedforward control 8, 10, 11
final control element 4, 6, 8, 11, 70–71, 73, 77–78, 83, 91
flow loop 11, 13, 19, 82, 93, 134
FOUNDATION™ Fieldbus 9, 69

gain 16–17, 19, 21, 47, 70, 76, 82, 87
grounding 25, 29, 40, 47, 49, 93

hand-held communicators 95
hardware 27
Harris index 15
HART® 9, 68–69
high-speed data recording device 65
histograms 73
historical data 34

I/O 27, 29
I/O check 27
incorrect mode 98
indicated variable 116
inherent characteristic 17, 19
input filtering 61
installed characteristic 17–18
integrating 74

key performance indicators 92
kickoff meeting 27

Lambda tuning 76
limited control 99
linear inherent characteristic 18
loop
    checking 1–3, 5, 8–10, 17, 19, 25, 27, 30, 39, 47, 78
    classification 115
    inspection data sheet 63
    performance monitoring 97
    performance program 91
    points 33
    rating system 92
    sheets 26
    tracking 34, 36
    loop tuning 70, 76, 128

maintenance
    predictive 90
    preventive 90
    proactive 90
    reactive 89
manipulated variable 115

mean 12–14, 73
mode logic 34
modes 34–35, 37

nonlinearities 16–17, 21

OPC (OLE for Process Control) 62, 97, 106
open loop 70
    bump tests 70
operator interface 34
oscillation 16, 73

P&IDs 26
PC-based simulation 27, 31
PID algorithm 6, 119
power spectrum 73
predictive maintenance 90
preventive maintenance 90
proactive maintenance 90
process
    conditions 34
    dynamics 70
    gain 74
    variable 4, 8
programmable logic controller 6
proportional, integral and derivative 6, 16

reactive maintenance 89
reproducibility 5
robustness 73

SAMA (Scientific Apparatus Makers Association) 26, 40
scan times 7
self-regulating 74
set point 4, 118
simulation 38
site acceptance test 8, 25, 39, 47
smart instruments 39, 40, 48, 50–51, 54, 93, 107
smart valves 103

standard deviation 12–13, 73, 98

test plan 26, 28, 40–41, 48, 52, 54, 61, 63
time constant 70, 74
time series data 69
total probable error 5
tracking 34–37
tuning 2, 6, 16, 50, 59, 62, 68, 73, 76, 80–83, 87, 89, 91, 97, 115

uncertain input 99
unit operations 38
units 33, 37–38

valve
    performance 50, 59, 89, 101–102
    predictive maintenance tool 102
variability 1–2, 11–14, 21, 73, 79, 89, 98–99
    index 15
    rating 14

walk-through 63
wire shielding 29

Ziegler-Nichols tuning method
    closed loop 133
    open-loop 132